Skyway

UNIVERSITY PRESS OF FLORIDA

Florida A&M University, Tallahassee
Florida Atlantic University, Boca Raton
Florida Gulf Coast University, Ft. Myers
Florida International University, Miami
Florida State University, Tallahassee
New College of Florida, Sarasota
University of Central Florida, Orlando
University of Florida, Gainesville
University of North Florida, Jacksonville
University of South Florida, Tampa
University of West Florida, Pensacola

SKYWAY

The True Story of Tampa Bay's Signature Bridge and the Man Who Brought It Down

Bill DeYoung

University Press of Florida
Gainesville · Tallahassee · Tampa · Boca Raton
Pensacola · Orlando · Miami · Jacksonville · Ft. Myers · Sarasota

Copyright 2013 by Bill DeYoung
All rights reserved
Printed in the United States of America on acid-free paper

This book may be available in an electronic edition.

21 20 19 18 17 16 6 5 4 3 2 1

First cloth printing, 2013
First paperback printing, 2016

Library of Congress Cataloging-in-Publication Data
DeYoung, William Leonard, 1958–
Skyway : the true story of Tampa Bay's signature bridge and the man who brought it down / Bill DeYoung.
p. cm.
Includes bibliographical references and index.
ISBN 978-0-8130-4491-0 (cloth: alk. paper)
ISBN 978-0-8130-6297-6 (pbk.)
1. Sunshine Skyway Bridge (Fla.)—History. 2. Bridges—Accidents—Florida—History. 3. Marine accidents—Florida—History. 4. Bridge failures—Florida—Tampa Bay—History. 5. Tampa Bay (Fla.)—History. 6. Lerro, John. I. Title.
TG25.T35D49 2013
363.12'30975965—dc23 2013015082

University Press of Florida
15 Northwest 15th Street
Gainesville, FL 32611-2079
http://www.upf.com

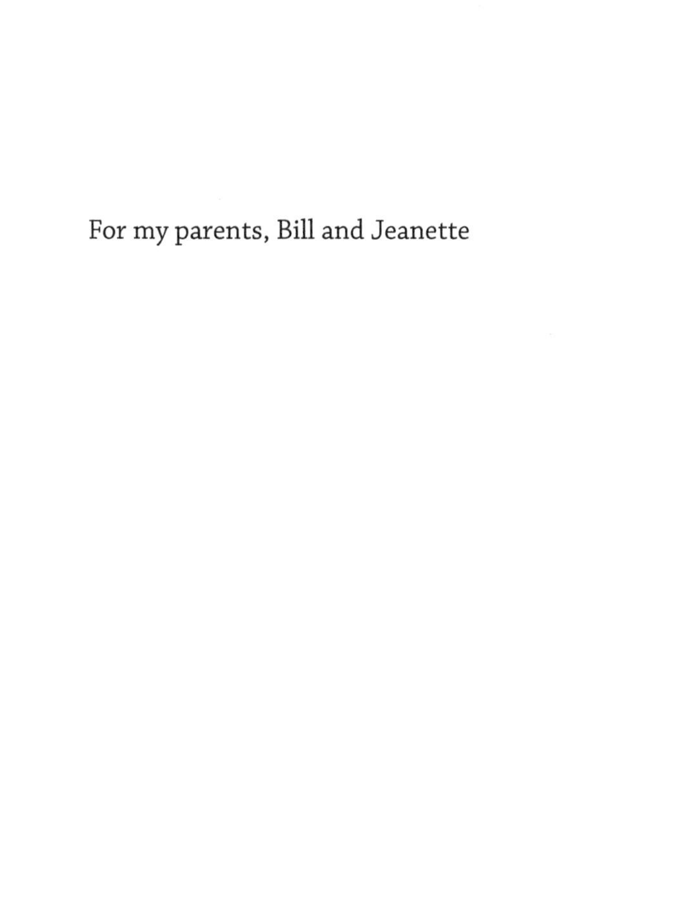

For my parents, Bill and Jeanette

See the dark night has come down on us
The world is living in its dream
But now we know that we can wake up from this sleep
And set out on the journey. Find a ship to take us on the way.
The time has come to trust that guiding light
And leaving all the rest behind,
we'll take the road that leads down to the waterside
And set out on the journey. Find a ship to take us on the way.
And we'll sail out on the water, yes we'll feel the seas roll
Yes we'll meet out on the water, where are all strangers are known.
If you travel blindly, if you fall
The truth is there to set you free
And when your heart can see just one thing in this life
We'll set out on the journey. Find a ship to take us on the way.
And we'll sail out on the water, yes we'll feel the seas roll
Yes we'll meet out on the water, where are all strangers are known.

Gerry Rafferty, ***The Ark***

Contents

Preface

There are no mountains in Florida. As a matter of fact, there's very little terrain you could even call hilly, except perhaps the gentle slopes of green and grassy horse country in the north-central part of the state or the bosomy ranges of sand dunes along the Atlantic Coast. No, Florida is flat, and it's wide, and because so much of it is only a few feet above sea level, the sky seems a million white miles high and a billion blue gallons deep.

In the early 1950s, before construction of the Sunshine Skyway Bridge, you could stand at the edge of lower Tampa Bay, at the tip of the Pinellas County peninsula, and gaze across fifteen miles of sparkling aquamarine water, where flocking pelicans dive-bombed into the waves and the silver scales of leaping tarpon reflected an afternoon's brilliant sunlight.

On a clear day, and with a good squint, you could just make out the treeline and water towers of Manatee County and try to imagine what in the world those people—whoever they might be—were up to down there. It was a different world, across the bay.

The Skyway was built, in the name of progress and expediency, to shorten the distance between these two growing metropolitan areas. Local and state government were interested, too, in a highway that would deliver tourists farther south, to spend their recreational dollars on the only Florida coast that did not have world-famous Miami Beach at its sandy terminus.

Much has been written about the adverse effect of unchecked growth and the population explosion in Florida, how tourism ultimately bit its feeding hand by destroying most of what was unique about the state in the first place.

And there's an argument to be made that constructing a fifteen-mile bridge across Tampa Bay—in the process interrupting one of the busiest shipping channels in the United States—wasn't such a great idea, regardless of its higher commercial purpose or its value as a photogenic landmark, a totem of tourism in an area that had nothing more spectacular to advertise than the sun and the sand and a whole lot of old people.

That's not the point of this book.

The Travel Channel lists the Sunshine Skyway at no. 3 among "The World's Top Ten Bridges," and residents of Florida's middle-western coast are rightfully proud of the majestic cable-stayed bridge, the fifth-largest such structure on the planet. Its bright yellow cables, set against the azure blue of the bay waters, or pink-and-orange Gulf sunsets, make for stunning landscape paintings and art photographs suitable for framing. It has indeed become an iconic Florida image, these days emblazoned on everything from T-shirts and coffee mugs to expensive limited-edition porcelain collectors' plates.

In 2012 the United States Postal Service put the Sunshine Skyway on a Priority Mail stamp, using a digital illustration by Chicago artist Dan Cosgrove.

Today there is little in St. Petersburg, or Bradenton, or any of the Tampa Bay cities, to explain that this beloved structure, this world-renowned marvel of engineering, is in fact the *second* Sunshine Skyway.

For a little over thirty-four years the opposite sides of the bay were linked by a bridge that was a state-of-the-art structure by the standards of the year it went up (1954). It too was considered an engineering marvel, and its unique design was proudly festooned on all the Florida souvenirs of the day, along with the requisite orange blossoms, flamingos, alligators, and water-skiing bathing beauties.

A booming population, and a related desire to make the area Interstate Highway–ready, resulted in the quickie construction of an adjacent twin Skyway in 1971. This happened to be the same year as the gates opened at Walt Disney World, which would radically change the face of Florida tourism.

And the population continued to swell.

By the end of that busy decade, however, red flags had been raised over the Skyway's construction and its conspicuous lack of protection against the unending procession of shipping vessels that passed beneath it every day of the year.

When the freighter *Summit Venture* struck the nine-year-old twin during a freak storm in 1980, the lamentations, recriminations, and hostile finger pointing became moot. Out of horrible necessity, yesterday's potential and what-ifs were made irrelevant.

* * *

I grew up in St. Petersburg and made my first Skyway crossing as a tot, in the back seat of my father's blue and white 1956 Chevrolet.

It was awesome. And, particularly for a little kid, it was terrifying.

At either end of the Skyway, in Pinellas or Manatee County, the driver's journey began with an innocuous five-mile ride over a series of small drawbridges and filled causeways, with tree-lined "rest areas" and shady picnic spots, broken by sandy stretches and mangrove stands with foot access for fishermen who enjoyed a rubber-boot wade into the waist-deep bay. Once you got on the over-water stretch of road, however, the picture changed dramatically. The Skyway was a ribbon floating tenuously on the blue-green surface, a lone thin highway with no buildings, no billboards, nothing to divert attention for the slightest instant.

As you moved along, in either direction, down those two puny lanes of asphalt, your eyes were drawn to the gigantic monster that seemed to rise up unnaturally in your path, arching out of the flat landscape toward the clouds, inevitable, immovable and getting closer all the time. You couldn't help but stare at it.

My earliest memory of that painfully drawn-out approach is of the crown of superstructure, the silver trestlework that sat atop the already impossibly high roadway.

It looked like a roller-coaster. Do we have to ride on that, I asked my father, and coast up and down on top of those terrible metal rises and falls that seem to scrape the clouds?

We didn't, of course, but the Skyway *was* very much like a roller-coaster. That 5 degree ascent, to which photographs don't do justice, was like the inevitable clack-clack up the slow climb of a coaster. You were committed. The anticipation and dread swelled in the back of your throat.

You'd reach the top—swallowed by the yawning maw of the silver superstructure—and have only a couple of seconds to admire the view before your downward dive began.

Even as an adult, I was always relieved when the road flattened out again, and I passed the sign that said I had made it to Manatee County.

And there were no lights on the bridge.

Now, I think about Wes MacIntire, and how his heart dropped into his stomach as the Skyway collapsed underneath him.

I think about John Lerro, who knew, as it was happening, that ramming the Skyway was the worst thing that could happen to a harbor pilot on Tampa Bay.

And I wonder about the thirty-five people who lost their lives that morning, and what must have been going through their minds. Were they dreading the roller-coaster, too?

SKYWAY

1

Ambush

John Lerro knew he was in trouble.

Squinting out through the slanted wheelhouse window of the freighter MV *Summit Venture*, the deputy harbor pilot couldn't see anything. Not the five loading cranes that stood at mute attention along the empty deck, which stretched out for five hundred slick feet of ship in front of him, five stories below where he stood.

Not the grumbling Chinese lookouts in heavy rain gear he'd posted to the bow of the heaving, 19,734-ton vessel less than fifteen minutes before.

Certainly not the lighted buoys tossing in the churning Tampa Bay waters, the markers that would tell him when it was time to turn this massive, unwieldy pile of steel awkwardly to port, in order to stay inside the invisible lanes of the shipping channel.

He couldn't see the Sunshine Skyway Bridge, two thin ribbons of concrete, silver steel, and asphalt that crossed fifteen miles of the open bay and rose to more than 150 feet at their pinnacle—easy enough to pass under with a good day's weather, but a target that required a sharpshooter's eye when an unanticipated squall like this one turned everything into unfocused black and white and terrible saltwater boil.

But John Lerro knew the bridge was there, in front of him somewhere.

It was 7:30 in the morning, May 9, 1980. A Friday. The sun had been straining to peek out from behind the ominous gray fog for almost an hour.

The gale had begun as hard, steady raindrops falling in rough rhythm from an opaque sky; within seconds *Summit Venture* was being suffocated by blowing white water coming from every direction at once, howling like a pack of hungry wolves trying to sniff out an opening to roar through. This was predatory weather, and it had pounced on *Summit Venture* at the worst possible moment.

As a veteran merchant mariner who held a full captain's license, Lerro had been in hard and sudden blows before. He knew that keeping calm, steadying his twitching nerves, was essential.

In his three years with the Tampa Bay Pilots Association, he had gone under that bridge almost a thousand times, safely guiding mammoth shipping vessels from all over the globe through the tricky inland channels and into—or out of—the Port of Tampa and the smaller ports in Pinellas and Manatee counties.

Flying a Liberian flag, the four-year-old *Summit Venture*—forged in a Japanese shipyard, 606 feet long and 80 feet abeam—had a thirty-man crew, all of them Chinese. As federal and state maritime laws mandated, the captain had consigned his vessel to the Tampa Bay pilot group to reach the port, where the holds were to be filled with pulverized phosphate rock. That morning it was John Lerro's turn in the pilot rotation.

He was thirty-seven years old, the youngest pilot in the organization. A third-generation Italian American from the Bronx, Lerro stood out among the rugged southern men who had been navigating the bay for decades and the leather-skinned tugboat captains who'd been hired as pilots through the back-slapping recommendations of their longtime friends in the group.

Resentment toward Lerro ran deep among the veteran pilots. Not because he was a fast-talking, opinionated New Yorker, with the wavy

Fig. 1. MV *Summit Venture*, 20,000 gross tons and 606 feet in length, known as a "handysize bulker" and built in Nagasaki, Japan in 1976. On May 9, 1980, flying a Liberian flag, it was to take on a load of pulverized phosphate at the Port of Tampa. Courtesy of www.swiss-ships.org.

black hair, deep-set brown eyes, and square jaw of a movie star. Not because he liked to talk about classical music and ballet.

Not because he was aloof or unfriendly—on the contrary, Lerro had a quick wit and a generous nature.

Many of the other men didn't like him because of a 1975 law that took the hiring of new members out of the pilots' hands and placed it under Florida's Board of Professional Regulation. Lerro had been the first pilot hired by the state body under what was, in effect, affirmative action. He wasn't one of them. He had been *forced* on them.

Taking *Summit Venture* into port was Lerro's only assignment for May 9. He was to close on a loan in the afternoon, in a Tampa bank, and use the money to buy stock in the Tampa Bay Pilots Association. After three years as a deputy he had earned the right to become a full pilot, share in the profits, and more than double his $40,000 annual salary.

He had an observer on board. Bruce Atkins—like Lerro—had decided to change his career course and become a harbor pilot. Even though the thirty-two-year-old Atkins had been a ship's master with Gulf Oil for years, he still had to spend thirty days as a "trainee" on Tampa Bay, riding with each of the association's pilots and deputies on routine transits, familiarizing himself with the bay. Atkins didn't want to be on *Summit Venture* on May 9—it was the thirtieth day, and he felt more than ready to begin piloting on his own.

Besides, Lerro had been his pilot during many Gulf Oil transits into Tampa, and Atkins didn't think much of the wiry New Yorker's handling technique. On more than one occasion Atkins had come within seconds of taking back control of a ship, as he just didn't think Lerro was making the right decisions.

"Ah, what the hell," he'd finally decided. It was just one more trip up the bay.

Once docked in Tampa, the crew—aided by the longshoremen at the port—would fill *Summit Venture's* five cargo holds, a three- or four-hour procedure. Then the next pilot in rotation on the Tampa side would negotiate the ship back out of the bay, and disembark at the Egmont Key pilot station, before the big freighter sailed for South Korea to unload at the Port of Pusan.

There was, of course, no need for harbor pilots on the open ocean. The sea is a wide highway with no designated lanes. No shoals, no tides or tricky currents to be concerned with.

Captain Liu Hsuing Chu had picked up Lerro and Atkins at 6:20 that morning, anticipating a routine transit up the roughly forty miles of water to the assigned port terminal.

At the Sea Buoy, before he'd started in, Liu had ordered the ballast tanks emptied. Normally the tanks are filled with seawater, to make an unladen ship ride more heavily—and steadily—on the sea. Discharging the ballast was standard procedure before an empty vessel was brought into port to load up. In the shipping business, time is money, and the real estate at the loading docks came with a high price tag: the sooner *Summit Venture* got in, the sooner it would get out, and another ship could put in at Rockport Terminal to start the whole process all over again.

* * *

As 7:30 approached, Lerro understood implicitly that his options were limited. The number one priority, for every one of the association's eighteen senior pilots and six deputies, was *don't hit the Skyway.* He was well versed in Tampa Bay's spider web of dredged deep-water channels and the treacherous shallows that could leave a ship stranded and helpless.

Both spans of the magnificent traffic bridge were crowned by complex configurations of silver steel girders that resembled the superstructure atop an old-time railroad trestle, or something a boy in the 1950s might build with his Erector Set. This distinctive feature, which made the Sunshine Skyway one of the most instantly recognizable tall bridges in America, was integral in holding the separate sections of cantilever arch together.

The black storm materialized when *Summit Venture* was a little less than a mile to the west of the Skyway. Monitoring the ship's radar, Atkins had just announced that he'd spotted buoys 1A and 2A, marking the tricky dogleg turn into Cut A—18 degrees to port—that would keep *Summit Venture* in the shipping channel and deliver it safely under the bridge's highest point before continuing to Tampa. They were almost there.

In that instant the curtain dropped. Visibility was zero. Blowing white water assaulted the thick windows and small wheelhouse portholes, dimly illuminated for a second at a time as if someone were flicking a light switch on and off in an adjacent room.

Bursts of static from the marine radio, crackling with the hard weather, punctuated the stale air in the small, white-paneled wheelhouse, competing with the baying wind and explosions of chest-rattling thunder for the attention of the five men on duty, each standing tense and hard-knuckled at his station.

Captain Liu and his crewmen looked nervously at Lerro, who stood fixed at the window, raising and lowering his binoculars, and at Atkins, whose eyes never left the starboard radar screen. Atkins was watching not only for the turn buoys but for the bridge itself, which would appear as a straight yellow line.

To Atkins's horror, as the radar made another sweep the center of the screen became solid yellow—damnable interference from the weather cell that had descended on them without warning.

Where was the turn?

Lerro raced between the radar and the window, the window and the radar, and began to consider his options. The Sunshine Skyway Bridge was up there, a fixed object, with Friday morning rush hour traffic poking along in both directions.

He issued orders for the bow lookouts to report sighting of a buoy on the starboard side—that would be 2A—and to make the anchors ready, just in case something drastic was necessary.

Like all big ships, *Summit Venture* had its own unique handling characteristics. But the mathematical logic was indisputable: a 20,000-ton ship, light in ballast and proceeding at half-ahead (about 9.5 mph) requires half a mile to come to a complete stop.

Turning hard to port, reversing back across the width of the channel, was not an option. Pilot Jack Schiffmacher was outbound with the empty gasoline tanker *Pure Oil*, Lerro knew, and was likely to be in the channel and coming his way just as *Summit Venture* was swinging around. A collision with a fuel ship could cause a massive explosion that could cripple or destroy both vessels.

A hard turn to starboard might be the best course of action, although with the wind out of the southwest, at his stern, Lerro would then risk having his ship—riding high with less than a dozen feet of draft in front—blowing broadside into the bridge. And the immediate depth of the spoil area, where the bay bottom sludge dredged from the channel had been deposited, was unknown. In the best-case scenario, *Summit Venture* would ground in the mud and wait out the storm.

The risk would be great, whichever way he turned.

A cry from Atkins broke his concentration. The radar had cleared, for one crucial sweep. "It's all right," Atkins said, his calm, professional voice belying the urgency of the situation. "I have the buoys. We're in the channel."

Just as quickly, on the next sweep, the yellow clutter returned, and the buoys were gone again.

The rain blew harder. *Summit Venture* plowed ahead.

Over the crackling onboard communication circuit, Captain Liu heard from the bow lookout. A turn buoy had been sighted.

Immediately Liu relayed this to Lerro, whose steely eyes were fixed on the window, as if staring hard enough would reveal the crucial turn markers' whereabouts through the darkness and wet ferocity.

"Where, captain, where?" Lerro replied, without turning his head. "I have to know!"

He didn't wait for the answer.

Summit Venture was now two-tenths of a mile from the Skyway Bridge, with no radar and no visuals. Because it takes time for large, heavy ships to respond to their mechanical commands, pilots have to think fast and plan their movements in advance, even under the best of conditions. Once Lerro understood that one of the Cut A buoys had been sighted, he chose his course.

There was, he reasoned in those tense seconds, no other choice but to assume he was in the right place. Any hesitation would bring him closer to certain disaster. With the wind behind him and a favorable current, *Summit Venture* should still be able to make the turn, and glide under the bridge, without incident.

Attempting to stop the ship now would unquestionably result in a collision with the Skyway.

"It's the best thing I've got," Lerro said under his breath. He gritted his teeth. He would "shoot for the hole," as the pilots called it. Atkins's brief radar sighting, and the report from the lookout, would have to do.

He ordered the engine speed reduced to Slow Ahead, and the 18-degree turn to port, to guide the ship into Cut A and safely between the towering spans of the Sunshine Skyway, through the 800-foot gap where the shipping lane flowed uninterrupted beneath the high roadway.

No one—not the bosun and carpenter, on lookout duty on the bow, nor the chief officer on his way forward to join them, nor the four men in the wheelhouse, pressed anxiously against the glass—could actually see the Skyway.

Summit Venture began easing into its turn.

The radio was nothing but static, so Lerro did not hear a call from Schiffmacher, in command of *Pure Oil*, advising that he was pulling out of the channel, east of the spans and well clear of *Summit Venture*, and anchoring in the shallows to wait for the vicious weather cell to beat itself to death.

Then the weather began to clear, as suddenly as it had materialized. Lerro brought up his binoculars.

From the starboard side, where the sun was making its first tentative appearance, he saw the Skyway. But it was not the center, not the 150-foot vertical clearance he expected. There was no accommodating gap. No hole to shoot for.

He saw that they were indeed making the slow turn but were hundreds of feet off course, south of the channel, and headed directly for one of the smaller concrete bridge supports.

The swirling wind, which had been blowing in gusts up to 70 mph, had changed direction in an instant, and now it was coming hard from the northwest, pushing the high-riding *Summit Venture* southward—and out of the channel. The high-riding freighter had "crabbed," or moved erratically sideways, without any of the men realizing it or the primitive detection instruments at their fingertips showing it. *Summit Venture* was a 20,000-ton balloon, floating on the water and at the mercy of a fickle and unforgiving wind.

Barely eight minutes had passed since the blackness and violence had descended. From his position in front of the radar station, Atkins peered forward.

It's the wrong set of abutments, he thought. *We're two to the right . . . OK, can we make it underneath that?*

Atkins had a mariner's moment of clarity. *If we give it a little more speed, so we could really steer . . . could we make it?* he said to himself. *We'll probably shear off the radar mast.*

But Lerro knew.

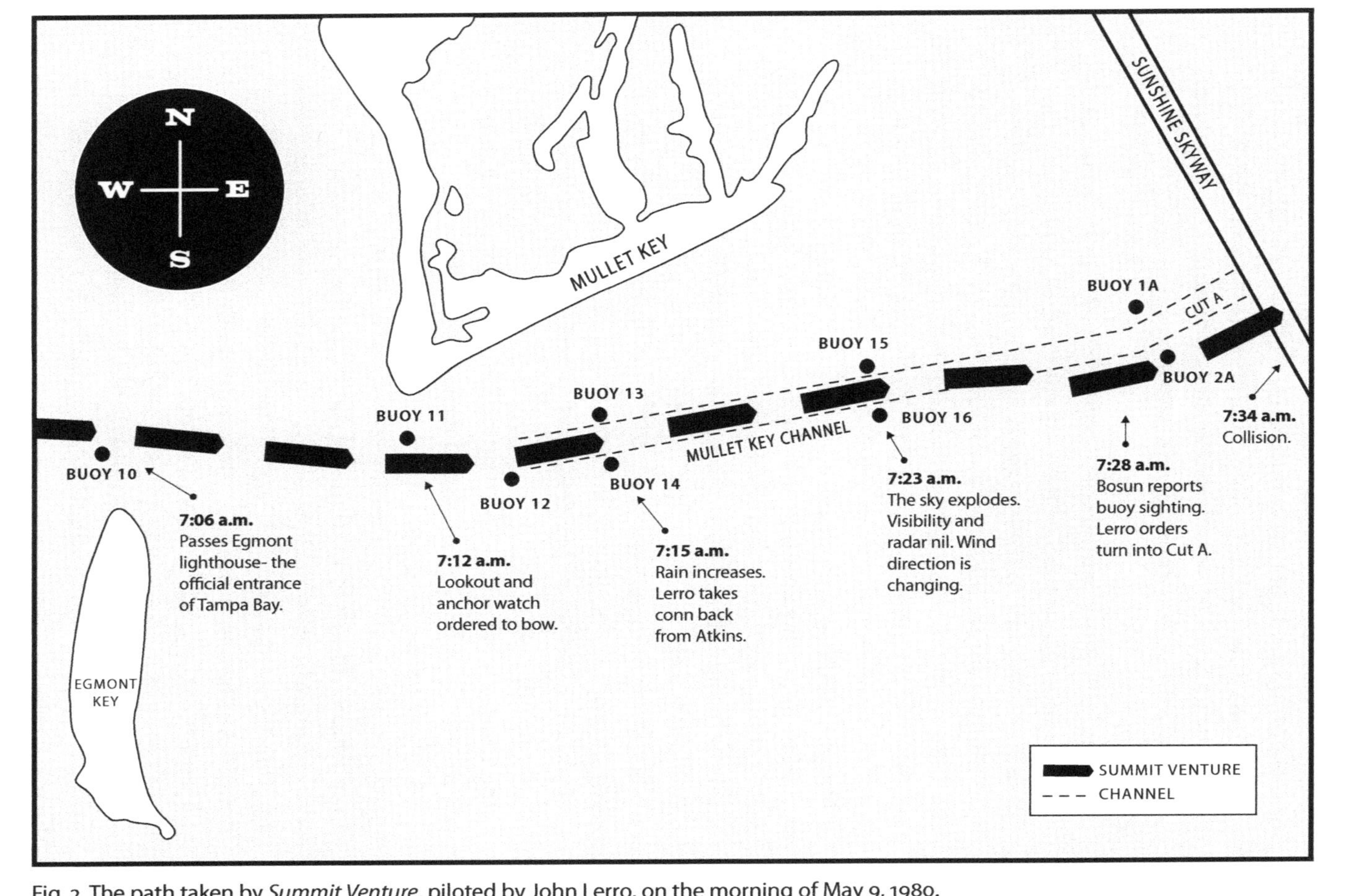

Fig. 2. The path taken by *Summit Venture,* piloted by John Lerro, on the morning of May 9, 1980.

Without waiting to give the order, he grabbed the engine telegraph and threw it astern twice—commanding the diesel engine to go in reverse. "Come hard port, drop both anchors," he shouted.

Ship's carpenter Lok Lin Ming, standing forward on the foc'sle, with his head down to keep the stinging rain from his eyes, had just made ready the port anchor. The order to drop came through, and he let it go.

At 7:34 a.m. the starboard bow of *Summit Venture* grazed 2S, the second-highest support pier of the Sunshine Skyway Bridge's Southbound span, and sheared off both concrete columns 54 feet above the waterline.

Lok felt the anchor hit bottom, and as he braked the heavy chain to hold the anchor fast, the Skyway—which he had just noticed for the first time, like a long-legged giant emerging from the mist in a cheap horror movie—began to groan, and twist, and fall.

Next to him, bosun Sit Hau Po, the lookout who had spotted the turn buoy, raised his eyes and saw the columns just before the collision. He didn't have time to drop the starboard anchor.

Terrified, Sit flattened himself between the iron anchor windlasses on the foc'sle deck.

When it began to rain steel and concrete, Lok leapt over the foc'sle railing, landed on the slippery main deck, and ran for his life.

From the wheelhouse 500 feet away, it was like watching a silent film in slow motion.

The impact, Atkins thought, was soft. *Pfft*, like hitting something with a sock. He didn't feel the vessel move or shudder.

The southern slope of the bridge came down in sections, one after another: a house of cards. The steep mountainside of road was stretched and rotated off its caps and piers; the old-style steel trestlework groaned, bent, and fell away. The concrete, Lerro observed, was popping off in small chunks, exploding and flying everywhere. He thought it looked like cornflakes pouring out of a box.

This can't be right, Atkins thought. *We didn't even hit it! How could we have caused all that devastation? What the [hell]?*

They all watched it collapse and tumble, piece by grinding piece, each man praying, wishing hard, silently pitting his own will against the inevitable.

And there were cars on the bridge, innocent people making their way south. Lerro, Atkins, Liu, and the helmsman stared up, stunned, immobile, as one vehicle after another flew from the break, nosed down and plunged 150 feet.

From their position behind the wheelhouse glass, nearly two football fields distant, they saw the splashes but didn't hear a thing. Then they would look up at the summit and watch another set of headlights appear at the jagged edge, point downward, and drop. The bridge was gone, but the cars were still falling.

The shaking pilot forced himself to grab for his radio.

Mayday, mayday, mayday, Coast Guard!

Responding at last to its most recent engine orders of full astern, *Summit Venture* began to groan in reverse. The port anchor was on the bottom, held fast by tons of shattered concrete, asphalt, and silver steel, and the chain pulled taut.

At least 50 feet of roadway, nearly intact, was draped across the raised foc'sle on *Summit Venture*'s bow, still bolted to mangled cantilever and trestle. Lerro looked at it, across the width of the vessel he'd been entrusted with, and marveled at the yellow-painted lines of the traffic lanes. The ship's forward mast protruded bravely from a hole it had poked through the falling road.

There were no cars or swimming people visible in the gray-green water far below the broken span. The men saw no wake, no ripple, no disturbance.

There was nothing.

The rain had all but stopped. The wind was gone. The sun was coming out.

Lerro was practically screaming.

Get all the emergency equipment out to the Skyway Bridge. Vessel just hit the Skyway Bridge. The Skyway Bridge is down! Get all emergency equipment out to the Skyway Bridge. The Skyway Bridge is down! This is a Mayday. Emergency situation.
Stop all the traffic on that Skyway Bridge!

2

Hands across the Bay

Pinellas County is a peninsula, jutting out and southward like an arthritic finger off the west coast of Florida into the Gulf of Mexico. It is the gateway to Tampa Bay, the largest open-water estuary in the state; in order for cargo ships to make the passage into the Port of Tampa, approximately forty miles west on the far side of the bay in Hillsborough County, they must pass through the fifteen-mile-wide strait between lower Pinellas and upper Manatee County. The bay covers nearly four hundred square miles, and the port is navigable via a labyrinth of man-made channels.

Since the early twentieth century, the Port of Tampa has been the busiest in Florida in terms of trade activity. Because the bay is shallow—the average depth is 12 feet—large ships with deep drafts must follow the channels precisely or risk grounding on the abundant shoals. Visiting ship captains can't be expected to familiarize themselves with these tricky passages, which are marked by a series of buoys, or to navigate them safely without doing damage to the ship, the crew, the cargo or the environment.

Vessel size and tonnage have steadily increased, and as Florida began to export as much cargo as it brought in—citrus, poultry, and phosphate fertilizer bound for the far corners of the world as well as for other American ports—so increased the demands on the Port of Tampa and the smaller port on the Manatee County side.

Before the Tampa Bay Pilots Association was established in 1886, incoming ships stopped at the lighthouse on Egmont Key—a 440-acre sand-and-scrub island at the entrance to the bay—to borrow a navigational chart from the lightkeeper. On their way out of port, the chart was returned.

The Egmont lighthouse was erected in the spring of 1848. That September it was destroyed by a hurricane—at one point, the entire island was under nine feet of water. The replacement, stronger, taller, and farther inland, debuted ten years later. It stands to this day, its beam broadcasting over the Gulf from 87 feet above ground.

Egmont's mixed history has included stints as an internment camp for Native Americans during the Seminole Wars and as a Union-occupied garrison during the Civil War (the better to stop Confederate supply ships trying to sneak into Tampa). At the outbreak of the Spanish-American War in the early twentieth century, it was home to Fort Dade, guns perpetually pointed toward the Gulf in anticipation of the Spanish invasion, which never came.

To the north of Egmont, just two miles across the water, is Mullet Key; here, on the southern-most tip of Pinellas, the Americans built Fort DeSoto, intended as Dade's twin post and similarly well armed with big guns and mortars. The bay's entrance was therefore fortified with heavy artillery, but never in history did it come under fire. The American guns were never fired in anger.

In 1912 the Tampa Bay Pilots Association leased two acres on Egmont, which by then was all but abandoned; Fort Dade was officially decommissioned nine years later. The association built a clapboard office and several small cottages for the pilots to get some sleep between assignments without having to be ferried back to the mainland. Ship-to-shore radio and telephone service arrived in due time, making communication between ship, shore, and pilot expedient.

They were specialists, experts in their chosen field. Local pilots who understood the bay and its delicate estuarine balance, and were trained not only in the safe operation of cargo ships but in docking at any of the port's berths, would board incoming vessels before they entered the bay. In Tampa Bay, as throughout the coastal United States, pilot service was compulsory for all foreign-flagged vessels, and for all U.S.-flagged vessels engaged in foreign trade, drawing seven feet of water or more. Pilots were required to hold a Coast Guard–issued federal license and a second pilot license from the state.

In a prearranged change of command, the pilot assumed control of the ship from the captain, guided it to the port, and safely docked it. After the ship's crew had loaded or unloaded their cargo, the next pilot's job was to get the vessel back out of port, across the bay and back into the open waters of the Gulf. There he disembarked—picked up by a small pilot boat—and returned to land to await his next assignment. The ships' operators paid handsomely for this service.

Transportation was a key issue in Florida politics in the 1920s. The speculative land boom would clearly not amount to much without good roads and a highway system. Governor John W. Martin began an ambitious project to connect every Floridian county seat via automobile thoroughfare.

In 1924 a trio of enterprising businessmen came up with the idea of ferrying passengers—and the few automobiles that existed in Pinellas County—across Lower Tampa Bay to the southern counties, Manatee, Sarasota and beyond. With a converted Mississippi River paddleboat, the *Fred C. Doty,* the Bee Line Ferry opened for business.

Even though the big, steam-powered wheel had been switched out for a diesel engine and twin screws, *Doty*, with its flat, low stern, was unsuited to rough-water travel, and the bay passage usually concluded with crewmen flushing oily saltwater out of the undercarriage. Seasickness was to be expected on these choppy forty-minute adventures, which covered twenty nautical miles from Pinellas Point in south St. Petersburg (the largest and most populated city on the Pinellas peninsula) to Piney Point in upper Manatee County.

The situation improved when another Bee Line boat, *Pinellas*, was added four years later. Built in 1882, the vessel had plied the waters of the Delaware River as the *Wilmington*. The ancient *Doty* was sold for scrap. Two voluminous ferries—the *Manatee* in 1935 and the *Sarasota* in 1936—were added to the fleet as cars became larger and passengers more plentiful. The vessels averaged 140 feet in length, and 34 feet abeam, and could each carry 30 automobiles and 180 people. They ran only during daylight hours.

With the advent of World War II, Florida became home to numerous military training centers. The armed forces took over Bee Line operations—more specifically, the government shut the ferry service down by requisitioning its boats. The *Pinellas* and *Manatee* were sent around the Florida peninsula to shuttle U.S. Navy personnel from Jacksonville to other strategic points along the east coast, while the U.S. Army used the *Sarasota*—which had originally been a cargo vessel servicing the Tampa-to-Everglades City route—for hauling troops from MacDill Air Force base, at Tampa's southern edge, up the bay into the city proper.

After the war the City of St. Petersburg's Port Authority resumed Bee Line service, adding a fourth ferry, a former Navy landing ship renamed *Hillsborough*. The population exploded in the postwar years as ex-military personnel—many of whom had spent months, even years, training in Florida—made the decision to bid farewell to the frigid New England or midwestern climate with which they had grown up and make a fresh start in the land of sunshine and oranges.

The idea of a bridge over Lower Tampa Bay had first been broached in 1924 by a local doctor and civic booster, Herman Simmonds. He worked hard to convince local and state government that he wasn't a crackpot—as some seriously suspected—and that the fifteen miles of open water at the bay mouth could successfully be spanned with a spectacular high-level suspension bridge over the Tampa ship channel.

Simmonds managed to get an act of Congress authorizing such a bridge and even obtained construction permits. But in 1927 the Florida land boom went bust; two years later the Great Depression enveloped the nation like a dark, wet blanket. Herman Simmonds and the Tampa Bay bridge were forgotten as government focused on more pressing matters.

It wasn't until 1948, with progressive thoughts once again running toward expansion, that the Pinellas County Commission began to entertain thoughts of connecting the two sides of the bay. During the Bee Line's enforced absence, motorists from the Pinellas side on their way to the undeveloped points south had been forced to travel over the aging Gandy Bridge, which crossed the northern sector of the bay, and around the outskirts of Tampa via U.S. 41. This was a detour of nearly fifty miles and a tremendous inconvenience, particularly with wartime gas and tire rationing.

The commission introduced legislation for a combination bridge and tunnel across the lower bay. The tunnel would not only be impervious to hurricanes and the effects of other bad weather; it would be strategically sunk far beneath the main ship channel into the Port of Tampa, and thereby would not interfere with the area's all-important import-export trade. Understandably, the Tampa Bay Pilots Association vigorously supported the tunnel idea.

It was discovered, however, that the proposed center of the tunnel—as the centerpiece of a series of bridges and causeways emanating from the Pinellas and Manatee sides—was technically in a tiny section of Hillsborough County. Since this proposal had not been advertised in Hillsborough, the whole thing was declared unconstitutional. Promoters were back to square one.

Tampa and St. Petersburg had always had an uneasy relationship. A center of industry in central Florida, Tampa was the bustling urban heart of massive Hillsborough County. Across the bay in Pinellas, St. Pete was viewed as a resort town, a place where people escaped the real world to fish, sit on the beach, or otherwise enjoy a life of leisure. And St. Petersburg attracted the elderly, who enjoyed the thought of living out their retirement years in a warm, sunny climate. The city was dismissively referred to in America's bigger cities (especially Tampa) as "God's waiting room."

During Florida's great Roaring Twenties land boom, St. Petersburg's civic leaders, businessmen and, significantly, newspaper editors began to advertise their community as "Florida's Sunshine City," attempting to lure prospective buyers with stories and images of the region's untapped beauty and the health benefits of its temperate climate.

Lew B. Brown, the flamboyant owner and editor of St. Petersburg's afternoon paper, the *Evening Independent*, declared—as a promotional gimmick—that every day the sun didn't peek out of the sky, for even a brief moment, the *Independent* would be given away free.

This practice continued for decades, even after the rival *St. Petersburg Times* bought the struggling paper in 1962. When the *Independent* was shuttered for good twenty-four years later, the bean counters announced that the paper had been given away just 296 times in its 76 years of existence.

* * *

After the war, when the St. Petersburg Port Authority bought the Bee Line ferry, engineers began to survey the lower bay to establish the feasibility of a Pinellas-to-Manatee crossing. They reported that the bridge would cost $8,627,000 and recommended a $10 million bond issue to pay for it. Immediately, the Hillsborough County Commission complained that the new bridge would "choke" Tampa and hamper navigation in the ship channel.

Eventually, the State of Florida took over all assets (and, importantly, debts) of the St. Petersburg Port Authority, including the submerged lands it owned. The bridge was fast-tracked by Governor Fuller Warren and the Florida legislature, who decreed that the project would be overseen by the State Improvement Commission, and the bridge would be built under the auspices of the State Road Board.

Tampa's local government, naturally, warmed to this idea. The writing was on the wall. Progress was inevitable, and tourism was already the state's largest industry. More people, whether they preferred the big-city life in Tampa or the shuffleboard courts in St. Petersburg, meant more money. And the State intended to pay for the bridge. Warren had already been instrumental in getting U.S. 19, the "Gulf Coast Highway" that began in Pensacola, five hundred miles up the coast, pushed through to St. Petersburg.

New traffic surveys were conducted, and the most expedient plan, all agreed, was to start the bridge at Maximo Point in southern St. Petersburg, with a series of landfill causeways. Small bridges, totaling eleven miles, would cross the open, shallow water of Tampa Bay, then a tall span would cross the main shipping channel, and the Skyway would end in another series of filled causeways on the Manatee side, outside the town of Palmetto.

There the road would join U.S. 41, the main artery running south through Manatee, Sarasota, and the other counties south of the bay. Also known as the Tamiami Trail, this pioneering, 284-mile thoroughfare opened in 1928 and provided the only link between the two South Florida coasts by running from Tampa south through Bradenton, Sarasota, Fort Myers, and Naples and then east through the Everglades to Miami.

First, Tampa Bay needed to be tamed.

Florida engineer Freeman Horton is historically credited for the design of the bay bridge—he was the first to submit a detailed plan for its construction, in 1945. But ultimately the state hired the New York firm of Parsons, Brinckerhoff, Hall and MacDonald for the task. For the all-important center span, the engineers settled on a tried-and-true cantilever through-truss support structure, 2,800 feet in total length. This included an 854-foot gap between the highest support piers, to be raised on either side of the existing ship channel.

Cantilever bridges balance their components on towers called piers that act as pivot points. After the support piers are driven deep into the seabed, steel arms are attached to reach out horizontally. The center span is literally "suspended" by crane and lowered into place, supported on each side—in the case of the Skyway, on the north and south—by the extended cantilever arms. The trestlework, or superstructure, is bolted into place to hold it all together and then coated with tough, weather-resistant silver paint.

In order to accommodate the largest cargo ships of the era, the roadway crossing the bay would rise to 149.5 feet above the waterline. Underneath, Mullet Key Channel was dredged to 35 feet deep.

The Skyway's precipitous 5 degree incline meant that vehicles would climb from about 12 feet over the water to 150 feet in a jaw-dropping 30 to 45 seconds. After passing over the web of metal grating that joined the two cantilever arms like tersely gritted teeth—a five- or six-second visit to the summit—vehicles would then descend on the other side at the same dizzying rate.

Fig. 3. Some fifteen thousand cars drove across the Skyway for free, under textbook Florida skies, on opening day, September 6, 1954. Courtesy of Florida Archives.

It was to be the first bridge in the country constructed entirely of prestressed concrete, reinforced internally with steel rods.

On July 4, 1949, Governor Warren took to the water, along with Alfred McKethan, head of both the Improvement Commission and the Road Board, and several key legislators from Tallahassee and other west coast cities and towns. They were ferried to Egmont and Mullet keys and shown the shipping channel and the proposed site of the new bridge, and the entire project was explained in detail.

What it meant was progress. Opening up an artery connecting the cities on the Gulf Coast would make things easier—not to mention more engaging aesthetically—for tourists. Visitors were pouring into Florida during the postwar years, and the raw wetlands, swamps, and piney woods needed to be held in abeyance if people were to stay and enjoy the beaches and balmy weather. And spend money.

"That period in the '50s was a period of growth and optimism," said Bob Graham, who would be Florida's governor from 1979 to 1987 and then a United States senator until 2005. "It was also the beginnings of creating a state out of what had been a scattered set of urban and rural areas across the peninsula and the panhandle. And that included things like the Florida Turnpike, and the bridge over lower Tampa Bay. And early planning for the Interstate system." Graham was a sophomore at Miami High School when the first cars rolled across the Sunshine Skyway Bridge.

Dredging for the bridge got under way in October 1950, and more than 4 million cubic yards of sand were removed from the bay to form the causeways. At a height of more than fifteen stories, the bridge would be taller than any existing building in St. Petersburg. By the time the project was completed the total cost was $21,250,000.

Opening was set for September 6, 1954, after three days of celebration and jubilation over Labor Day weekend.

On March 15 the *St. Petersburg Times* announced a "Name the Bridge" contest, in partnership with the St. Pete Jaycees and the Junior Chamber of Commerce, and with the support of the State Road Board. Entries would be accepted for exactly one month. More than twenty thousand suggestions were received by the April 15 deadline, many accompanied by florid poems, sketches, and religious imagery.

The Jaycees judged them on five distinct points—"salability," "easy to remember," "descriptive," "geographic area" and "advertising value."

Numerous suggestions were rejected because they reinforced the widely held belief that St. Petersburg, now being aggressively promoted as "Florida's Sunshine City," was still pretty much a haven for the elderly, the convalescing, and the soon-to-die. The Jaycees whittled the list down to twenty finalists, which were then submitted to a panel of journalists, advertising executives, and travel agents. Eleven of the top twenty used the prefix "Sun."

The name Sunshine Skyway was contributed by Indian Rocks Beach businesswoman Virginia Seymour, who owned and operated the eight-cottage Gulf Ranch Motel with her husband. On September 3, she was a guest of honor at the Dedication Ball, at the St. Petersburg Coliseum, where she received an engraved plaque, a wristwatch, and a framed painting of the Skyway.

At the September 4 dedication ceremony five thousand people—believed to be the largest crowd yet assembled in St. Petersburg—heard former governor Warren praise Seymour's "remarkable and fascinating combination of words, a name with peculiar and mellifluous charm."

Warren also lavished praise on *St. Petersburg Times* owner, editor, and publisher Nelson Poynter. "No one has spent more man hours on behalf of this Skyway—he was more devoted and consecrated to it than anyone," the former chief executive said as the diminutive newspaperman with the important voice smiled sheepishly. "Almost from the day I took office, he started hounding me for this project and kept at it."

The ceremony was held at Al Lang Field, the city's minor league baseball park. Other speakers included U.S. Senator Spessard Holland, General James A. Van Fleet (recently retired; he talked about a hopeful, prosperous future "with little likelihood of an Atomic World War III"), incoming governor LeRoy Collins, and acting governor Charley E. Johns. As the highest-ranking Florida official, Johns was to cut the ribbon when the bridge was officially opened on September 6.

Johns had been president of the State Senate and had become acting governor upon the death in office of Daniel T. McCarty—Warren's successor—in September of 1953. He was now a lame duck, having been defeated in the ensuing special election by his popular and charismatic fellow Democrat Collins, who had twice been voted "Most Valuable Senator" by the Capital Press Corps in Tallahassee. The two had a frosty relationship.

September 5 was a Sunday, and in the evening an all-faith religious ceremony was held at Al Lang Field. After a rousing performance by the St. Petersburg High School Band, the assembly was attendees treated to three sacred songs by blues singer and actress Edith Wilson, who played Aunt Jemima on radio and television commercials. Wilson sang while dressed in her character's checkered dress and apron, wearing the stereotypical white "Mammy" scarf on her head.

The evening's keynote speaker was Bishop William C. Martin, president of the National Council of Churches of Christ in America. A Texas-based Methodist preacher, Martin delivered a speech called "Bridges of Faith in a Divided World," in which he compared the construction of bridges to man's "instinctive urge to span the chasms that have separated him from his neighbors." The parallel, of course, was "in building bridges of human understanding." He patiently explained this fact in case it hadn't been obvious.

Across the bay in Manatee County the president of the Bradenton Chamber of Commerce presided over a day of water-skiing contests and exhibitions on the beach by "Texas" Jim Mitchell, who ran a Sarasota reptile farm and miniature circus. The Miss Sunshine Skyway bathing beauty contest was followed by dinner and a floor show at the Bradenton Yacht Club. Sixteen-year-old Sonja Opp, of St. Petersburg, was crowned.

Excitement was in the air on both sides of Tampa Bay, as the eyes of the country—and, perhaps, the world—were about to turn to the area for the very first time.

On opening day, September 6, the *St. Petersburg Times*—the staunchest and most influential supporter of the bridge project since its origins before the war—published a 278-page special edition under the front-page headline "Hands Across the Bay." Article after article praised the Sunshine Skyway's innovative construction and

Sunshine Skyway Souvenir Edition

St. Petersburg Times

Sunshine Skyway Souvenir Edition

At Long Last

HANDS ACROSS THE BAY

Here Is Guide To Your Sunshine Skyway Edition

It's Classified . . . . .

TOLL FREE UNTIL 11 P.M.

Skyway Opens At 10 A.M. Today

Stickers, Badges

Button Holders

Fig. 4. Equal parts news, boosterism, and wishful thinking, the special opening day edition of the *St. Petersburg Times* was a carefully built marketing tool. It contained fifteen separate sections and weighed four pounds. Courtesy of *Tampa Bay Times*.

trumpeted its importance to the economic future of what would now be officially known as the Sun Coast.

"The Sun Coast Cities have now joined hands," wrote Poynter. "They meet as Good Neighbors and Good Partners in a common enterprise—the simple but gigantic job of providing the most attractive area in the world in which to live and visit."

In his editorial Poynter went on to take a good-natured jab at his competition, the city that had balked and harrumphed at the first mention of a lower-bay bridge. "The center of gravity has moved from Tampa to the coastal cities," he wrote. "It is not that Tampa has diminished but merely that tourism and light industry of the coastal cities have grown faster than the needs for heavy industry and commerce."

Consisting of fifteen separate sections and weighing four pounds, "Hands Across the Bay" was the most ambitious printing project Poynter and his staff had ever undertaken. It cost 35 cents at newsstands. The initial run of 112,000 copies, the *Times* boastfully reported, was not enough to satisfy public demand, so the presses were plated to print a second batch.

They were not all snapped up by local residents eager to commemorate the historic day, however. "Copies of the big edition went to every Associated Press editor in the United States," the *Times* announced, "inviting them to prolong their stay in St. Petersburg and at the Gulf Beaches when they come to Tampa in November for their annual convention." Equal parts news, boosterism, and wishful thinking, "Hands Across the Bay" was a carefully built marketing tool. Poynter was doing his part to keep the money flowing.

Poynter's special edition was stuffed column-to-column with advertisements for the "resort" towns to the south—Bradenton and Palmetto in Manatee County, then Venice, then Sarasota—and for the barrier islands on the Gulf side of St. Petersburg, which were now christened the "Holiday Isles."

A letter from President Eisenhower, addressed to Acting Governor Johns, was dutifully reproduced inside: "Through the addition of this fine bridge, the new Gulf Coast road will serve to make the beautiful resorts of your State most accessible to thousands of visitors traveling by automobile, and I am sure that the bridge will be a valuable addition to our national highway program."

So endorsed by the country's highest power, the word was out. Thanks to this sleek, attractive new bridge, the vibrant, exciting west coast of Florida was open for business.

* * *

Despite the Weather Bureau's prediction of overcast skies and scattered thundershowers, dedication day dawned under a dome of blue. Dressed in his trademark vested pinstripe suit, Johns made a speech at the crowded ceremonies in front of the Maximo Point toll booth on the Pinellas side. In his brief comments the acting governor called the Skyway "a mighty and majestic monument to the cooperative spirit of man."

Fig. 5. Opening day on the Pinellas side: Florida's acting governor Charley E. Johns and Miss Greece 1954, Rika Diallina, with "Miss Sunshine Skyway" bathing beauties representing the ten counties newly connected by the bridge. Courtesy of Florida Archives.

Ten coastal counties were represented by pretty young women in red swimsuits and pointed silver crowns. In a large, primary-colored jigsaw map erected behind the speakers' podium, each in turn slid a geographically correct wooden representation of her home county into place. The final piece, an oversized cartoon cutout of the Sunshine Skyway Bridge, was presented to Johns by Rika Diallina, the current Miss Greece, who had recently competed in the Miss Universe pageant in California, coming in at sixteenth place.

During Senator Joe McCarthy's red scare, Diallina had initially been denied entry into the United States because of alleged communist ties. She reportedly obtained a special visa through the intervention of Secretary of State John Foster Dulles, arriving in Long Beach only days before the Miss Universe pageant. Her replacement, Effie Androulakakis, bowed out of the final round of competition in order to give Diallina her day.

Johns snapped the last piece into the map, symbolically linking the ten counties. A cheer went up, and the throng broke into applause.

Also on the makeshift stage was James Melton, who had grown up in North Florida, had spent the years 1942 to 1950 as one of the Metropolitan Opera's leading tenors, and had had a musical variety TV program, *Ford Festival Presents the James Melton Show*, for two seasons on NBC. With his stage career winding down, he'd recently turned his collection of antique cars into a museum for tourists in Hypoluxo in southeastern Florida. At the ceremony at Al Lang Field, Melton had done the honors on "The Star-Spangled Banner."

Johns cut the ribbon, photographs were taken, and he took his place in a motorcade across the new bridge. At the front of the impatient line was a rebuilt 1900 Rockwell hansom cab, with Miss Greece in the passenger seat and Melton behind the wheel. Holland and Van Fleet rode in the second car, followed by the other dignitaries.

At the precise moment the motorcade reached the Skyway's zenith, the Bee Line boat *Manatee*, 150 feet below, struck her flag, symbolizing the passing of a torch and the end of an era.

In Palmetto on the Manatee side, more speeches were made. Instead of a cut-out map, long violet ribbons represented the ten linking counties, collected by Miss Greece from the same bathing beauties,

Fig. 6. Opening ceremonies at the Manatee County tollbooths. From giant papier mache hands, the bathing beauties unspooled symbolic ribbons, one for each county. These were delivered to Miss Greece, who then "presented" them to Acting Governor Johns. Courtesy of *Tampa Bay Times*.

who arrived astride giant, linking papier-mache hands. Diallina then brought the ribbons to the acting governor, solemnly proclaiming, "I present to you the united bonds of the State of Florida."

Fuller Warren, in his speech, told the throng that "it's worth the $1.75 just for the view from that height." The former governor was effusive in his praise for the bridge's designers and builders, and the politicians whose vision made it all possible, and for the "good neighbor policy" that had "always" existed between Pinellas and Manatee. "Now there can never be an Iron Curtain between these two counties," he pronounced.

Shortly after 11:00 a.m. the Sunshine Skyway Bridge was opened to motorists, who could cross the bay for free, as many times as they liked, until 11:00 that night. More than five hundred cars waited on the road outside the Pinellas-side toll booth, with nearly as many on the southern end.

When the $1.75 per vehicle toll went into effect, a quick tally showed that slightly more than fifteen thousand automobiles had crossed the Skyway in its first twelve hours.

The tolls, the *Times* had reported in "Hands Across the Bay," would be collected until the funds reached $22 million, in order to pay off bond money. The bridge would therefore eventually pay for itself, and after that, passage would not cost drivers a thing.

Meanwhile, the State continued to collect Skyway tolls. In 1959 the fee dropped from $1.75 to $1. The "mighty and majestic monument to the cooperative spirit of man" would be declared obsolete and shuttered for good a little over thirty-two years later. At no time after that first day was it ever a free bridge.

3

Warning Signs

On September 6, the very same day Florida and Pinellas County were congratulating themselves on their progressive nature and their engineering brilliance, University of Florida sociology professor T. Lynn Smith was in Rome, Italy, addressing a United Nations Conference on population. St. Petersburg, Florida, Smith told the assembly, was the leading "retirement city" in the United States, and cities were replacing small towns as America's "old folks' homes."

Between 1950 and 1960 Florida's population nearly doubled. Families from the north, weary of the bitter winters, were setting up house, starting up businesses and changing the societal landscape. When the 1960 census was tallied, Florida was the tenth most populous state in America.

By 1965 tourism in the bay counties was at an all-time high, and it was understood that the steady tide was going to swell into a tsunami once Walt Disney moved in. This was going to be a game changer in terms of tourism, intra-Florida commerce, and the overall economic health of the state.

Fig. 7. In the 1950s the ubiquitous "Mr. Sun"—who had a bespectacled, grandfatherly appearance—invited visitors from the southern counties to visit St. Petersburg via the Skyway. Courtesy of *Tampa Bay Times.*

Disney had quietly begun buying up scrub and swampland in Central Florida in 1964, using dummy corporations to keep his name out of the transactions and keep the price down. The public knew nothing about it until the *Orlando Sentinel* broke the news on October 20, 1965. Disney quickly called a press conference and admitted that yes, he had purchased 27,433 acres for his "Florida Project," five times the size of his original theme park in Anaheim, California. He'd paid about $180 per acre.

Two months later Florida's governor Haydon W. Burns—who had personally introduced Walt Disney at the November press conference—proposed a refinancing of the Sunshine Skyway project, with the ultimate goal of four-laning the bridge by constructing a second two-lane span identical and adjacent to the 1954 edition. An aging two-lane bridge was not sufficient for the traffic flow of the mid-1960s, and when Disney World opened outside Orlando, Florida was going to have to be able to handle the exponential increase. And because the crucial lower bay crossing would then be up to speed for the Interstate Highway system, the Skyway would be eligible for federal money.

In 1960 Interstate 75 had been extended across the upper bay, from Tampa to northern Pinellas County, via the Howard Franklin Bridge. Burns's proposal would extend it southward through St. Petersburg as I-275, across the Skyway, and into Manatee and Sarasota counties.

Burns's $56 million "three-point program" therefore included providing direct, expedient access from the southern counties, the Skyway, and St. Petersburg to Interstate 4, which was already being extended into Tampa to connect with Interstate 75. Interstate 4 was a direct link to Orlando and Walt Disney World.

St. Petersburg Times publisher Nelson Poynter, while he agreed that the four-laning was necessary in the name of progress, publicly criticized Burns's campaign to pay for the expansion via another expensive bond issue.

Under the 1954 terms, the bonds that had built the original bridge were to be fully paid by 1968, which would mean the end of tolls—the Skyway would finally be a free road, as promised. Editorialized the *Evening Independent*, now also owned by Poynter: "The St. Petersburg City Council's endorsement of the plan is typical of the unseemly rush with which this proposal is being railroaded to completion. Instead of rubber-stamping this pre-election scheme of Governor Burns, the council should be concentrating on seeing to it that the state lives up to its commitment to make this a free bridge as soon as the present tolls are retired."

The governor, however, got his way, and the four necessary state agencies (Road Board, Development Commission, Board of Administration, and Bond Review Board) approved the bond issue. Under this agreement the Skyway would continue to collect tolls for another thirty years, until 1996.

To his credit, Burns lowered the $1-per-car toll to 50 cents. It should

Fig. 8. In 1966 Governor Haydon Burns announced a plan to build a "twin" Sunshine Skyway via bond issue. The plan was wildly unpopular with the public, who had been promised a toll-free bridge once the original bonds were paid off. Reducing the tolls was the governor's compromise; the second bridge would open five years later. Courtesy of *Tampa Bay Times*.

be easily understood, the governor explained, that "you can't funnel the traffic of U.S. 19, and the traffic of Interstate 4, through expressway facilities and then expect a two-lane bridge to accommodate them."

Using the 1954 blueprints designed by Parsons Brinckerhoff, the state began construction on the second Skyway—to the west of the original, with about 100 feet between them—at the end of 1967. Opening was penciled in for late 1969.

A large crack was discovered in pier 1S, the southernmost of the new main piers (1N, its mate, was apparently unflawed). In an April 1969 inspection of 1S by divers, twenty-two additional smaller cracks were found. Something had gone wrong. The project groaned to a halt.

The Florida Department of Transportation (as the Road Board was renamed at the end of the 1960s) hired J. E. Greiner and Company, independent consulting engineers, to see what was going on.

Maintenance engineer J. F. Fernandez wrote a letter to his supervisor in Tallahassee:

> As seen from my viewpoint, maintenance inherits built-in troubles that will surely increase with time and the addition of dead load, wind load, and impact load whenever the structure is completed. In my opinion, if there is any possibility this pier can be removed and replaced it would be a savings to the taxpayers as well as saving much unwanted publicity and embarrassment in the future. To replace the pier now might cost two million dollars, but to have to replace it 5, 10 or 15 years from now would cost an unheard-of price and should be considered a catastrophe.

According to Greiner, the state's DOT engineers had used bottom-boring data from 1954 in evaluating the best drilling sites and procedures for the second bridge.

This meant that the foundation pilings for pier 1S had been driven into sand and not bedrock, and in another cost-cutting measure, concrete had been used for these pilings, instead of much stronger steel, as recommended. During construction of the outer frame of the tall pier, this crucial base support had shifted. Pier 1S tilted more than one inch, opening up a half-inch-wide crack in the concrete frame, 20 feet in length. The consultants termed the shallow piling embedment "poor practice" and the subsequent shift of pier 1S "intolerable."

Driving steel supports into the substrate and realigning and resetting the pier added $3 million and two years to the project. While repairs were being made, Deputy State Highway Engineer William Gartner Jr. received a memo from state structural engineer T. Alberdi:

> As requested by you, the following is a written explanation of why additional subsurface information was not obtained for the new Sunshine Skyway project:
>
> The time factor for preparing the construction plans for this project was extremely tight. It was at the end of Governor Burns' administration and our office was pressed for time to prepare plans for the four-laning project. Apparently, it was felt by the Department that the existing soil information would be adequate.

The "twin" Skyway was dedicated by Governor Reubin Askew on May 19, 1971, with much less fanfare than its predecessor. Indeed, there was a "ho-hum" factor involved: in 1954 the Skyway was wondrous. This second one was merely functional.

Built to the west, on the Gulf side of Tampa Bay, the new bridge carried vehicles from north to south, while its older sibling was assigned northbound traffic from the south counties. Small green-and-white signs bolted to the steel girders overhead informed motorists going each way they were on the W. E. "Bill" Dean Bridge. The former chief engineer for the State Road Board, Dean had helped to build the original Skyway—and by proxy, its modern double. A pioneer in the use of prestressed concrete in bridges, Dean died in 1965, three years after retiring from the Road Board. No one ever referred to the Skyway as the W. E. "Bill" Dean Bridge, though, and the little signs at the top of the spans were the only clue, historically, that he had been thus honored.

Under state law, DOT inspections of the both bridges were carried out every two years. The 1974 inspection revealed a series of small underwater cracks in pier 1N, the new bridge's *other* tall pier, which DOT engineer Jack Roberts described to the media as the sort of "settlement" they had expected. There was, he proclaimed, "No reason to be alarmed." But eyebrows were naturally being raised. As the *Sarasota Herald-Tribune* editorialized:

> It would seem there should be an investigation begun, from the top, into the whole story of the engineering and construction of the second Skyway project. Whoever is to blame should be identified and held responsible. If the problem is the result of negligence on the part of people on the state payroll, they should be penalized. If the cause turns out to be shoddy work by outside engineering or construction firms, they should be required to make restitution and disqualified from doing state work. To do less is to invite more scandalous public work blundering.

In May 1976 the agency reported cracking in all of the newer bridge's thirty-two supporting piers. Caused, according to DOT, by a chemical

Fig. 9. The ill-fated twin Sunshine Skyway span was opened in February 1971. It would last a little over nine years. Courtesy of Florida Archives.

reaction that occurred soon after the concrete had been poured in the 1960s, they began as "micro-fractures" that were exacerbated by heat, time, weather, and the intrusion of salt water. Roberts said his team found "no evidence" that any of the all-important supports had shifted, and that there was no danger to the structural integrity of the bridge, or to motorists.

Both Skyways, meanwhile, were taking it from all sides. Every year small boats—fishing vessels and privately owned pleasure craft—struck the various smaller bridges on occasion, resulting in emergency lane closures and repair work. A drifting barge caused $35,000 in damage in 1973. Four years later a trawler rammed one of the lower-level support piers and nearly cut the roadway in half.

Then there were the larger issues: in May 1978 harbor pilot John "Jack" Schiffmacher was bringing the 851-foot *Phosphore Conveyer* out of the Port of Tampa, at 10 knots, when a sudden loss of steering left the ship on a collision course for the main spans. Schiffmacher ordered the anchor dropped, and *Phosphore Conveyer*'s forward momentum was slowed as thirteen shots of chain—more than 1,000 feet—ran out and the anchor dragged the muddy bottom. The pilot, the captain, and the wheelhouse crew all tensed.

The vessel groaned to a stop 40 feet from the Sunshine Skyway.

Meanwhile, the Skyway itself was a money machine. The economic growth predicted in the 1950s had exceeded anyone's expectations, and by 1978—twenty-five years after the first bridge went up—an estimated 72 million vehicles had made the crossing, paying $51 million in tolls. This averaged out to more than 5 million vehicles per year.

Concerns about the safety of the twin bridges, and their vulnerability to a ship strike, were raised in a *St. Petersburg Times* article published in the wake of the *Phosphore Conveyer* incident. Under the headline "What Could Make the Skyway Bridge Fall Down?" writer Don North outlined the results of the most recent DOT report on the condition of both the 1954 and 1971 spans.

After the discovery of considerable cracking in many of the *older* bridge's support piers, which resulted in continuous exposure of the inner steel rods to the weather and corrosive sea air (an effect called spalling), in 1978 the DOT officially listed the original span in "poor" condition, with a sufficiency rating of 51.1. Anything below 50 would require federal money for a replacement. The second span, barely seven years old, had been rated "fair," with a sufficiency rating of 61.

The larger and more immediate issue, North pointed out, was the conspicuous lack of "fenders": fixed, sturdy obstacles installed around the main piers to stop or deflect wayward vessels. Each roadway support pier had a concrete "crash wall," commensurate with the total width of the pier, from which two concrete columns rose and connected with the steel truss and the roadway. The crash walls of the tall piers rose 15 feet above the waterline, enough concrete to deflect small boats but an ineffectual line of defense against a cargo ship weighing thousands of tons.

In 1954 the state had installed a wall of wooden flatboards lashed to pilings alongside the channel under the Skyway; these had long ago rotted away, and an even smaller cluster of wooden poles, bound with rusty cable, was all that existed to "protect" the new span from any out-of-control waterborne monster that might come blundering through.

"How far should we go in being our brother's keeper?" asked Roberts, by now the DOT's chief of bridge maintenance, in the story. "Should we put armor plating over houses to protect them from airplanes? Shouldn't the ships themselves have some sort of backup systems? They go through lots of bridges."

Each Skyway span was designed to support loads up to 80,000 pounds and to hold firm even in the most intense and dastardly weather conditions, including hurricane-force winds. But their high piers were vulnerable, Roberts admitted, to a lateral strike from a moving vessel.

"I'd say an empty ship could do it," he said. "Some of them are very heavy."

Concluded reporter North: "What has never been officially conceded, apparently, is the fact that loss of a pier would bring the whole metal skeleton, roadway and all, crashing down. If any of three piers on either side of the main channel are similarly hit, there will be a $32 million splash."

Privately, Roberts was frustrated by the inspection reports he was getting. The "new" bridge supports, he knew very well, should not be cracking so seriously, and so quickly. In a 1979 memo Roberts asked the DOT budget office for money to hire an independent engineering firm to study—and repair—the extensive cracking. Since 1974, he explained, "we have been unable to determine the seriousness of the problem. While physical monitoring has indicated the structure is stable, this does not reveal when the deterioration may progress to the point the structural integrity becomes a question."

The Roberts memo, along with others highlighting DOT's private concerns over the safety of the bridge, were brought to the attention of the media in the spring of 1980, in the wake of the *Summit Venture* disaster. The whistleblower was retired civil engineer Arthur Goodale, a construction superintendent for the company that had been contracted to build the 1954 structure.

He wasn't involved in any way with the second span, but Goodale had been keeping tabs on the bridge, and he believed the extensive cracking was due to DOT's widespread use of "cheaper" materials in the concrete mixing process, among other factors. He had the paperwork to prove it.

After the *Summit Venture* incident, Goodale made it a personal crusade to make sure the public knew that the Sunshine Skyway had been, in his words, "an accident waiting to happen."

4

The Combat Zone

The 1980s dawned inauspiciously. On Tampa Bay, it was business as usual. Shipping was an around-the-clock industry, and the association's pilots worked twenty-four hours a day, seven days a week. There were eighteen of them in January, along with six deputy pilots, who with each Gulf-to-port and port-to-Gulf passage took another step toward a more advanced license and the responsibility and right to handle larger ships with deeper drafts (the section of hull that rode beneath the water). The salaries, naturally, got better too.

According to Port of Tampa records, shipping on the bay increased 60 percent between 1970 and 1980. In 1979 the Tampa Bay Pilots Association moved 4,974 vessels, over 500 more than the previous year.

On February 9 DOT received a report from Howard Needles Tammen and Bergendoff, the independent consulting firm hired at Roberts's insistence to evaluate the troubling cracks in the 1971 bridge. Of the thirty-two support piers, thirteen were found to have serious cracks, some an eighth-inch wide or more. Still, repairs were not made.

* * *

The Coast Guard cutter *Blackthorn* was based in Galveston, Texas, where its primary responsibility was to maintain and repair buoys and other aids to navigation in the complex waterways of the coastal Gulf.

Since October 1979 *Blackthorn* had been in drydock in Tampa. Built in 1943, the 180-foot diesel-electric buoy tender had been around the world dozens of times on various service missions, most of them routine navigational assignments and/or rescue operations. The Coast Guard had sent the ship to Tampa for a complete overhaul—her age and constant activity necessitated improvements every few years.

In Tampa *Blackthorn*'s main propulsion generators were rebuilt, and a large section of the shell plating was replaced on the port side. The sanitation system was vastly improved (something the fifty-man crew found particularly gratifying). By January 28, 1980, which happened to be the sixty-fifth anniversary of the U.S. Coast Guard, the repairs were finished, and the young crew reported back for duty.

The commanding officer was thirty-four-year-old George J. Sepel. After his men had conducted the necessary tests of radio and navigational equipment at the dock, Sepel took *Blackthorn* onto the bay for a two-hour sea trial.

This usually routine exercise revealed a problem with the main power generator, which the crew was unable to fix. Back to port went the vessel; the malfunction was quickly repaired, and the crew settled in for the voyage back to Galveston. Just before sunset, at 6:04 p.m.—several hours behind schedule—the *Blackthorn* began its journey down Tampa Bay toward the open Gulf.

Coast Guard vessels, like tugs, fishing boats, or recreational watercraft, did not require a licensed harbor pilot. Sepel would later testify that he wouldn't have known how to hire one even if it had been required. Neither Sepel nor any of his deck officers had navigated Tampa Bay prior to *Blackthorn*'s inbound journey four months earlier. Seven members of the crew, in fact, were new to the Coast Guard and had never sailed before.

The sky was clear that evening, without fog or rain to hamper navigation. It was a relatively balmy 61 degrees. *Blackthorn* was fully loaded with 51,270 gallons of fresh water and 26,695 gallons of diesel fuel. Also aboard were five motorcycles, a small car, three motorbikes, and a refrigerator, items picked up during the crew's long layover in Tampa, destined for homes and families once the passage back to Texas had been completed.

Tampa Bay pilot Gary Maddox was conning (piloting) the Soviet-owned passenger cruise ship *Kazakhstan*, coming out of a different berth at the Port of Tampa. Because his ship was operating at a faster speed than *Blackthorn*, Maddox radioed the cutter to request that it pull out of the channel and allow *Kazakhstan* to pass. This was a relatively easy maneuver, as *Blackthorn*'s draft was just 12 feet, so Sepel and his helmsman complied and let the larger vessel by, in Cuts C and D.

After returning to Mullet Key Channel, Sepel directed Ensign John "Randy" Ryan, his Officer of the Deck, to take the conn. The CO went out onto the port wing, just as the Skyway was looming overhead. *Blackthorn* was slightly to the right of the center of the channel.

Sepel peered out into the night, illuminated by a three-quarter moon, and saw another vessel moving inbound, about 2,500 yards away. Its navigation lights were just beginning to separate from the lights of *Kazakhstan*.

He dashed back into the wheelhouse. "Where the fuck did he come from?" he shouted.

Veteran pilot Gene Knight was on MV *Capricorn*, a 605-foot tanker carrying 151,611 barrels of crude oil from St. Croix in the Virgin Islands to the Florida Power dock at Weedon Island Station in Pinellas County.

Capricorn's loaded draft was 31.6 feet, perilously close to the dredged bottom. Knight kept her tightly confined to the channel. Seconds before, he had safely passed pilot Maddox on the outbound *Kazakhstan*, their meeting prearranged via VHS radio by mutual one-blast whistle signals. This meant a standard port-to-port passage would be observed.

Knight first saw *Blackthorn* as the smaller ship was passing outbound under the Sunshine Skyway, in Cut A. Pilots referred to this area as the Combat Zone because the Skyway was perilously positioned close to the 18-degree dogleg turn incoming vessels made from Mullet Key Channel into Cut A. Coming to a stop, particularly for the huge bulkers and tankers, was not an option so close to the Skyway. Ships inside the Combat Zone were at their most vulnerable and pilots were at their most vigilant here.

There was little room for error and no regard for luck. It was all about skill.

Knight attempted several times to contact *Blackthorn* by radio, and received no response. Still, it looked as though the normal port-to-port meeting was possible until the two ships were about 1,200 yards apart. At that point, Knight realized that the cutter was literally coming down the center line of the channel and would soon be on his side—in his precious space.

On Knight's order, *Capricorn* made a slight turn to port, then an evasive hard turn. He issued a two-blast whistle, indicating his intention to make an unorthodox starboard-to-starboard passing. Knight knew an emergency when he saw it.

"Right full rudder," shouted Sepel, effectively taking back the conn from Ryan. He had the engines put back full and ordered "Standby for collision" piped over the onboard sound system.

"That guy just kept turning to port, kept turning . . . turning," Sepel would later testify. "I could see a tremendous bow wake . . . it was like he was pushing a wall of snow."

At 8:21 Knight ordered a four-whistle blast—the danger signal. The men on the command bridge of *Blackthorn* felt the blood in their veins turn to ice.

The Coast Guardsmen at anchor watch had been idly watching the colored deck lights of the receding *Kazakhstan* and didn't even see *Capricorn* until she was directly in front of them. Three of the four sailors scrambled off the bow and onto the buoy deck. The seaman serving as "phone talker" (communication with the bridge) could not unhook his headset and was therefore trapped on the bow.

The vessels collided, virtually head-on. *Capricorn* was still into her hard turn to port—Knight was concerned that he was too close to the vulnerable Skyway to straighten her out—and *Blackthorn* scraped hard down the massive tanker's port side.

The terrible screech and low rumble immediately roused the rest of the crewmen, many of whom were reading or watching television belowdecks. *Capricorn*'s 13,500-pound anchor caught on *Blackthorn*'s hull and ripped the steel plating like aluminum foil. The massive hook lodged deep in the sailors' shower area.

As the ships continued to move, *Capricorn*'s anchor chain paid out and came to what sailors refer to as "the bitter end." This was the death blow. Almost immediately, seized at chain's length by the relentless forward motion of the 14,000-ton tanker, the wounded *Blackthorn* began to list to port. Sepel stopped the engines, but *Capricorn*, still in the port turn, only tightened her grip.

Sepel's last order before *Blackthorn* capsized was "Abandon ship!"

All was chaos aboard the cutter; the emergency lighting failed, and as the final top-to-bottom roll began—just twenty seconds after the first list to port—many of the inexperienced crewmen were standing in front of a bulletin board, double-checking their assigned stations in an emergency.

Fifteen men were trapped on the mess deck when the little vessel rolled over. Three forced open the watertight door on the starboard side and escaped.

Several crewmen later testified that they had no idea where the lifejackets were stored. Of the five inflatable life rafts aboard *Blackthorn*, none worked properly. One, it was discovered, had the word "BAD" painted on its case. It was missing the CO_2 cartridge that would inflate the raft automatically, and it had been stored for somebody else to deal with later.

The shrimper *Bayou* had been following the buoy tender out and was on the scene within minutes. Its crew picked up twenty-three men.

Rescued twenty minutes later by another Coast Guard vessel, four sailors had been holding afloat an unconscious shipmate, Chief Warrant Officer Jack Roberts, in hope that he could be saved. Roberts was dead when pulled from the water.

"I had seen the ship and was watching it close in," a crewman was overheard to say once he was safely aboard the *Bayou*. "I was waiting for an order to turn, and the order never came."

Blackthorn disappeared beneath the still waters of Tampa Bay four minutes after the collision; twenty-three men drowned—nearly half the crew. "You couldn't see the people in the water," ship's cook Robert J. Fitzgibbon testified, "but you could hear them screaming for help."

Divers from the nearby Eckerd College Search and Rescue Unit dove *Blackthorn* about two hours after the sinking and knocked on the hull in a vain attempt to find survivors in an air pocket.

Separate investigations by the National Transportation Safety Board and the U.S. Coast Guard agreed, for the most part, on the cause of the accident. Among the NTSB findings:

> CO Sepel's lack of recent seagoing experience, and his unfamiliarity with the bay, made his decision to sail at night imprudent;
> *Blackthorn*, with its light draft, could easily have stayed to the extreme edge of the channel—or left it altogether—to allow *Capricorn* full access to the deep water it required;
> Conceivably, the accident might have been avoided if the Coast Guard had hired a harbor pilot.

Among the additional blunders cited by the Coast Guard were Sepel's failure to keep fully aware of the situation and "effectively supervise his relatively inexperienced conning officer"; and Ryan's incomplete understanding of the use of whistle signals and the differences between the rules of the road in inland versus international waters.

The Coast Guard also assigned partial blame to George P. McShea, master of *Capricorn*, and pilot Knight for failing to post a proper lookout on the ship's bow and failing to sound the danger signal in a timely manner.

In the end in a proceeding called an Admiral's Mast, both Sepel and Ryan were issued letters of admonition—the lightest punishment possible—as permanent entries in their military records.

* * *

Shipping ground to a virtual standstill while the *Blackthorn* wreck remained where it had settled, partially blocking the wide Cut A channel underneath the Sunshine Skyway. An auxiliary channel was opened, and all vessels were required to maneuver carefully through an even

Fig. 10. Sailors from the Coast Guard cutter *Blackthorn* await medical attention following their rescue from Tampa Bay on January 28, 1980. Twenty-three crewmen died when *Blackthorn* collided with the tanker *Capricorn,* capsized, and sank in the Skyway shipping channel. Courtesy of *Tampa Bay Times.*

smaller opening under the bridge. Ships had to proceed so slowly that they barely had enough speed to maneuver. It was a test of nerves, even for the hardened harbor pilots. Nobody was happy about any of it.

On the afternoon of February 16, under clear skies, the 750-foot bulk carrier *Jonna Dan* was being piloted into port. The U.S.-flagged ship was the largest vessel the Coast Guard had allowed in since the *Blackthorn* accident. Once berthed in Tampa, it was to load up with cattle feed.

Negotiating the narrow bypass proved tricky for the pilot aboard *Jonna Dan.* He stopped the ship's forward motion just before passing under the Skyway, but the stern continued to swing to starboard. In seconds it had clipped pier 1N, one of the main channel piers holding up the southbound span. A chunk of concrete, ten feet long and two inches wide, chipped off and dropped into the bay. The pilot quickly maneuvered the freighter back into position, avoiding further damage. *Jonna Dan* was purposely grounded and subsequently refloated by tugs to proceed into Tampa for inspection and repairs.

The bridge was inspected and deemed relatively unscathed—repairs amounted to $40,000, which DOT called "minor." The Skyway would not have to be closed.

Ironically, *Jonna Dan* had destroyed the single remaining wooden piling, the useless "protector" of the southbound Skyway, which had been there since 1971.

The *Jonna Dan* pilot, as was customary, filed the accident report. The Board of Pilot Commissioners' investigator, retired Coast Guard Admiral Irwin Stephens, advised against taking any action against him—indeed, Stephens said, the incident might have been "catastrophic" were it not for the quick thinking of the pilot.

His name was John Eugene Lerro. He had been on the job for three years.

5

Captain Lerro

All ships move gracefully, slowly. A ballet dancer moves gracefully, always with follow-through motion. They have the same effect on me.

John Lerro

Although he'd been an active member of the Sea Scouts in his youth, John Eugene Lerro wasn't particularly interested in the water. Like most teenage boys growing up in the middle-income boroughs of New York in the 1950s, Lerro's primary interests were cars and girls. According to family legend, his dad had to twist his arm to get him into the maritime industry.

Charles Lerro was a civil engineer and project manager with New York's White Construction Company. His wife, Theresa, taught elementary school music and gave private piano lessons at the family home. Their daughter Julie was born in 1937; John arrived five years later, on October 20, 1942. In 1953 the Lerros moved from the Bronx to the Jackson Heights neighborhood of Queens.

Both John's grandmothers had emigrated from Italy, and raised their children as devout Catholics, but as adults Charles and Theresa were not regular churchgoers. Although Julie went to a public high school, John—at his own insistence—had a fully parochial education. The Lerros believed in art, music, and literature. They were smart, and they were cultured, and their modest home was filled with books, and paintings, and classical records.

"They expected me to go to college," Lerro said of his parents. "My mother used to tell me that the Romans once ruled the world, to remember that my heritage is Roman. But I was kind of a street urchin and didn't have serious thoughts at first. Where I grew up, it wasn't an Italian neighborhood. The kids—there were me—the Wop—two Irish kids and two German kids. And it was the Wop who got his nose bloodied most of the time."

In August 1960 seventeen-year-old John Lerro began life as a cadet at the State University of New York Maritime College. Founded in 1874, it was the very first maritime school in the country, and its mission was to train seamen and officers for the merchant marine while still giving them a "proper" education. Like the Navy, the shipping industry required competent, knowledgeable, and disciplined personnel. A life at sea was hard, but for many young New Yorkers of modest means, it was a way out of the boroughs. It was a way to see the world—and to make a good living at the same time.

The fifty-five-acre college was housed at Fort Schuyler, a French-style fortress built during the Civil War on the Throgs Neck peninsula, at the crucial intersection of the East River and Long Island Sound in the Bronx. Each summer, in order for cadets to gain more practical knowledge, there was a supervised cruise to Europe aboard *Empire State IV*, the school's somewhat antiquated 565-foot training ship.

On campus, cadets were housed in dormitories called compartments, twenty-four to a room, with six sets of bunk beds going up each side. In 1960 the new class of "Mugs" arrived to find the freshman compartments being repaired and refitted after a fire; *Empire State IV*, at anchor in its usual place beneath the Throgs Neck Bridge, was to be their temporary housing.

Alphabetically, the young cadets were assigned spaces in the ship's cramped, smelly cargo holds. Lerro's first roommate was Robert House; the two became fast friends and shared quarters at Fort Schuyler for all four years they attended the school. Eventually, as they gained seniority, the underclassmen were able to move off the

Fig. 11. Cadet John Lerro's Maritime College senior photo, June 1964. "His weekend activities include, among other things, going to the ballet, taking ballet lessons, and practicing at home." Courtesy of SUNY Maritime College.

ship and back onto terra firma. The dorm buildings were finally reopened in 1963.

John Lerro and Bob House were inseparable. Cadets were given precious little leave time—usually a Saturday night, if they hadn't received any demerits during the week—and more often than not John and Bob would go into the city together to have dinner or catch a movie. Lerro, with his wavy black hair and dark eyes, was extremely popular with the ladies. As his friends all knew, it was an image he carefully cultivated.

"He had a romantic, Italian flair," House recalled. "He loved beautiful women and good food, and he was very elegant and careful about his appearance. Very positive and sweet and romantic." John Hayes, who bunked in the room next door, remembered Lerro talking, only half-kidding, about one day trying his luck in Hollywood. "Meet the right people," he'd say with a wink, "and you're in like Flynn."

But the marine school was the only certainty in their young lives, and so they leapt out of bed every morning at 6:15 at the sound of reveille, showered, shaved, and dressed in the school work uniform, woolen shirts and trousers and navy blue caps. Cadets learned not to deviate from the routine. After breakfast came cleanup—issued mops, brooms, rags, and brass polish, they were required to scrub their living and working quarters to a military shine.

After inspection, it was off to class. The average workload included the usual college classes in mathematics, English, history, and science, alongside such make-ready subjects as navigation, naval law, marine engineering, and mechanics.

Although Lerro had enrolled at the academy at his father's insistence, it did not take long for him to realize he actually enjoyed the discipline and routine of the place, the dressing, spit-shining, marching, and inspections. The more he learned, the more adventurous stories he heard, the more he began to look forward to a career at sea, standing a midnight watch on a heaving deck somewhere on a distant ocean.

Back in Queens, Theresa Lerro died during her son's freshman year.

Fort Schuyler was justifiably proud of its athletics programs. House boxed (his friends affectionately called him "Boomer"), and Hayes ran cross-country. Lerro, perhaps because of his boyhood Sea Scout training, joined the college rowing team as a freshman. Instead of sailing smoothly like Ivy Leaguers in slender and delicate sculls, he worked with monomoy boats, heavy, double-ended wooden craft based on a classic whaleboat design. In the mid-twentieth century monomoys (also known as surfboats) were standard-issue lifeboats on Coast Guard and naval vessels. They were tough and they were sturdy.

Maneuvering an eight-man monomoy required muscle, patience, and a keen understanding of wave movement and tidal patterns. Lerro, who saw this sort of rowing as a way to stay in top physical condition, absolutely loved it. He was on the rowing team for all four years. He kept one of his broken oars on the wall of his dorm room and always bragged about the grudge-match regattas, where his team did battle with cadets from Kings Point Merchant Marine Academy.

Everyone looked forward to the summer training cruises, away from the classrooms and textbooks of the Fort, churning across the open ocean, getting invaluable practical experience in ship operations.

"He had a romantic notion about the sea," recalled Hayes. "I remember one night coming up on him, in our junior year, walking in the fog on deck. The whole thing seemed to excite him—the idea of ships in the fog, trying to find their way."

Then there were the ports. Marseilles, Lisbon, Liverpool, Copenhagen. On one trip *Empire State* docked in Naples, and Lerro and House continued to Rome, where sister Julie had emigrated after college. She had a job teaching English to Italian schoolchildren. John's father met them for a grand reunion. The cadets were awestruck by the art and history of Rome. Exploring foreign streets, said Hayes, "was part of the allure of going to sea."

It was a multifaceted allure. "We were in Dublin the week after President Kennedy had been there," House recalled. "John was really a good-looking guy, with that uniform and that officer's hat. The girls were all over him. We couldn't buy ourselves a drink in the bar there, because we were Americans and they loved Kennedy."

Back at school, John had discovered the impressive legacy of alumnus Edward Villella. Like Lerro, he was a handsome young man of Italian descent, athletic and masculine, but obsessed with the arts. Villella had also come from a Roman Catholic family in Queens. He was a tough, scrappy kid who'd been forced by his domineering mother to sit through his older sister's ballet classes; much to his parents' chagrin, Villella began to dance too. He liked it. And he was good at it.

As a teen, he met legendary choreographer George Balanchine at the School of Ballet Arts in Manhattan; encouraged, he dreamed of becoming part of Balanchine's stable of New York City Ballet dancers.

But Villella's father had other ideas. Ballet, he insisted, was not for men, and Eddy needed to travel a more masculine career path. Joseph Villella, a Garment District truck driver, insisted his son go to college, and he was paying. No one in the Villella family had ever gone to college, much less earned a degree.

Because his best friend's brother was attending SUNY Maritime College, Villella applied there to study marine transportation. He didn't care about marine transportation—indeed, he didn't even know what it was—but he liked the look of the place and the spectacle of the Saturday afternoon dress parades, so when the school accepted him, he accepted it. He figured it was as good a place as any.

Only too aware that the sexuality of a male ballet dancer would be called into question by his fellow students, Villella made it a point to participate in school athletics. "Instead of waiting for the taunts, I acted first, pounding out my frustrations on the heavy bag every chance I got," Villella wrote in his autobiography. "They'd all see what they'd have to deal with if they crossed me." Villella put his street smarts and natural scrap to good use and eventually became the school's welterweight boxing champion. He won a letter on the varsity baseball team.

During the summer cruises on *Empire State*, however, he'd lock himself in his cabin and practice plies and pirouettes. Villella graduated in 1959 with a degree in marine science.

The following year he became one of Balanchine's principal dancers with the New York City Ballet. He never put his degree to any practical use.

Lerro began to read everything in the college library about its most famous graduate. He asked the instructors, the staff, even the campus barber, what Villella was really like. "The thing that really interested him about Villella was that he was a very manly person," House said. "As opposed to your usual image of ballet dancers. He fought well above his weight class, and was a success, which was impressive to John."

During their senior year, Lerro, House, and Hayes took a discretionary course in art history. Once a week, on class field trips, they went into the city to visit the art museums. Lerro, who considered himself quite the connoisseur, would hold forth on artistic style and substance.

"You weren't supposed to leave the college during the week," House said. "But I had a car there, and it was possible to sneak out. It's called jumping ship. And if you got caught you'd have some problems, but it wasn't like a big, big deal. And we never got caught." On these illicit trips off campus, pugilist House—after dropping Lerro off at some museum, diner, or movie palace—would drive to Gleason's Gym for a workout in the ring.

One day, out of nowhere, Lerro changed the routine. "I was leaving and doing something, and so John got it in his head that *he* wanted to do something," House recalled. "So he went, on his own, to the Martha Graham Dance Studio. Now, Martha Graham's company had a practice studio right next to Carnegie Hall. And John just walked in the door and said he'd like to get into it."

According to House, John signed up for ballet lessons on a whim. He hadn't mentioned it before that day.

"He was tremendously excited by it, and he was apparently quite successful. They liked him a lot, and they said he had some beautiful movement. Apparently he would've gotten into the company and actually become a member of the Martha Graham Company, as opposed to just practicing. But it never happened. By the time he finished with college he was going into the merchant marine, so it just didn't work out. But he did it for two, three years."

One of Lerro's first dance teachers in the Carnegie Hall studio was a pretty young woman from Brooklyn named Sophie. Although he continued his ballet studies with other instructors, John began to see her socially. When he wasn't sneaking off campus to take another lesson, or attend a gallery opening, he was out on the town with his new love. They were both obsessed with dance, theater, and the other arts.

Lerro met his idol face to face several times. "A friend of mine, who was a college professor in New Jersey, put on concerts for the students, and he invited Villella and the prima ballerina Patricia McBride," said House. "Both of them were stars in the Balanchine ballet. I couldn't go for some reason, and John got the assignment of picking up Villella and McBride, and driving them to New Jersey. That was a big thrill for him. And when Villella would dance at Lincoln Center, John and I would go to the concerts and afterwards go backstage and meet him. And he was always very nice to us."

In his 1964 senior photo for *Eight Bells*, the college yearbook, John Lerro cut a dashing figure. In his dress white uniform and epaulets, with piercing dark eyes, strong Roman nose and chin, and heavy shock of wavy black Tony Curtis hair, he looked like a matinee idol. He looked like a guy who was going places.

Everyone in the senior class got to write his own biography for *Eight Bells*—after four years of tightly regimented schoolwork, drilled in day after day with military precision, the cocksure cadets were happy to inject a little fun into their parting shots. Lerro wrote:

> John was recommended to Fort Schuyler by the New York City Youth Board. He was given an alternative, Elmira or Fort Schuyler. He made the mistake of coming to the Fort. John has the distinction of being the only cadet in the history of the school who got four haircuts in four years. His weekend activities include, among other things, going to the ballet, taking ballet lessons, and practicing at home. John intends to ship out and if possible, continue his career in the field of dance and acting. He hopes to attend the Yale School of Drama.

Graduates received a diploma and a third mate's license.

* * *

As far as Lerro and his buddies were concerned, joining the United States Navy was not an option. Although graduating officers were routinely offered active commissions, by 1964 the war in Vietnam was beginning to heat up. "We all turned them down flat on the active commissions," said Hayes. "And the Navy retaliated by turning us all down flat on reserve commissions."

Gene Sweeney, one of Lerro's rowing teammates, said graduating from the academy as an officer was "a big thing to have in your back pocket. If you did get drafted, you could say, 'Thank you, I'm going to go on board a Navy ship as an officer.' Instead of crawling around in the swamps over there."

The Queens and Brooklyn draft boards made exceptions for seafarers; as long as they shipped out after graduation, and could provide proof of their employment, they would not have to go to war. "If you took all the mariners out of Brooklyn, for example, you wouldn't be able to man the merchant marine," Hayes said.

A car accident in his senior year damaged Lerro's knee and effectively ended his ballet career.

On graduation day, Hayes, always thinking one step ahead of the others, slipped away from the after-parties and made straight for the International Masters, Mates and Pilots Union, where he joined up—and immediately landed a job at sea.

When Lerro and House arrived at the union hall the following morning, proudly clutching their freshly minted third mates' licenses, Hayes was able to tell him he was already shipping out. His friends eagerly joined the union too. But business was slow on the high seas, and Lerro spent a lot of 1965 working eighteen-hour-a-day shoreside jobs, which he dismissed as "glorified slavery." He and House briefly shared a walk-up apartment on Ninety-Fifth Street, near Central Park.

In 1965 John and Sophie married and took up residence in the Inwood section of northern Manhattan. Their only child, a boy they named Charles Christian Lerro, arrived on January 13, 1966.

With the escalation of the war, qualified merchant mariners became more important than ever. Supplies had to be delivered, expediently, to the other side of the world. For seven months Hayes served as chief officer on an ammo/supply ship, SS *Oceanic Cloud*, sailing in and out of the war zone from nearby ports in Southeast Asia. By pulling a few strings, he was able to bring in Lerro as second officer, third in command.

Although both Hayes and House eventually left the merchant marine to pursue other careers, Lerro continued to sail. For a time he worked for Sea-Land, the company whose pioneering work in container shipping (as opposed to bulk cargo) made it indispensible to the U.S. Department of Defense during the Vietnam era. He obtained his master's license, which led to the command of foreign-flagged ships, bigger money, and even bigger adventures. Sophie continued her ballet career by teaching at Ballet Arts, New York's second-largest dance school.

"She is the most sensitive, spiritual and cerebral person I've ever known," Lerro would say. "She is a lot of superlatives. We both love beautiful things, art objects. My favorite is Louis Comfort Tiffany glass because he gave everything a form and a life."

Being away at sea for four or five months at a stretch left little time to appreciate the finer things, however. It was during the six years the family spent in Miami that Lerro began to consider a career switch to piloting. Not only was the money better and the work potentially less dangerous; piloting would keep him in one place and allow him to be home more or less every night to sleep in his own bed. And to be there for his young son.

In January 1976 Lerro was hired as a pilot in training by the Panama Canal Company, and the family relocated to Central America. They lived in Coco Solo, the former U.S. Navy submarine base near Colon, which the company used for civilian staff housing.

For Lerro, this was likely a transitional move—Panama was considered the ideal training ground for harbor pilots, as the canal's complex system of locks required the sort of precise maneuvering they needed to understand and master for the navigation of tricky American harbors.

Sophie taught ballet at the Panama Canal Zone Armed Forces YMCA. Young Charles made so many friends that he pitched a fit when his dad announced that they were moving back to the States, eleven months after they'd arrived in Panama.

During his time on the canal, Lerro also devoured the navigational charts of every one of Florida's shipping ports. His family liked Florida, and several of Lerro's best maritime buddies either worked there as pilots or made frequent visits along their shipping routes. It wouldn't be totally unfamiliar.

Lerro's timing could not have been better. The Florida Legislature, the previous summer, had passed a bill creating a Board of Pilot Commissioners. Under the auspices of the Department of Professional Regulation, the board would review, hire, and license harbor pilots in the state, in effect taking such matters out of the hands of the local pilot associations.

* * *

Judy Nunez was twenty-four years old in 1976. Her father, B. F. Wiltshire, had practically raised his three daughters on Egmont Key. “My dad would literally leave me on the island, and I’d stay at the house if I didn’t feel like going back to Tampa, over the summer,” Nunez said. “He would go next door and tell the pilots that he knew, and trusted, that I was going to be there. And they’d keep an eye on me.”

Wiltshire had been a Tampa Bay pilot since 1957, and as association manager in the mid- to late ’70s it fell to him to set the transit schedules as well as to arrange a working rotation. Each pilot was on duty for three weeks, and had to be on call and available day or night to move a ship, followed by three weeks off.

After twenty years on the pilot “bar” in Tampa, Wiltshire knew each man’s strengths and weaknesses. He knew the hard workers, the high and mighty, the tough talkers, and the rabble-rousers; he knew the ex-merchant mariners, the no-nonsense Navy vets, and the grizzled former tugboat captains. He knew the drinkers and the churchgoers, the loose-lipped and the stoic.

Wiltshire’s job required him to understand and accept the rivalries and tensions that existed among the pilots, and to assign each man’s duties accordingly.

“If you’re the boss, you put this group of workers on the shift because they get along very well together,” Nunez said. “Each side got along within their respective unit. But they didn’t like the other guys. There was an ‘us and them’ mentality for as long as I can remember. They weren’t politically motivated. They weren’t fighting for anything in particular. They just genuinely did not like each other.”

According to Bob Thompson, a veteran captain based in the Tampa area, pilots were quite adept at sizing one another up. “As a pilot, you know all the other pilots,” Thompson said. “So if you’re Unit 10, and you’re going to be meeting Unit 6 on the bay, you probably know the characteristics of that guy.

“You may say ‘I don’t want to meet that guy in the turn.’ So in that way, you get to know the styles of all the other pilots on the bay. That’s a very important part of being a pilot. You know who you’re meeting. You know what they’re doing.”

Thompson graduated from the SUNY Maritime College several years after Lerro, Hayes, and House. Like them he had gone to sea as a mate and eventually earned a master's license. Over the decades, they all became good friends. Much like airline pilots, harbor pilots consider themselves highly skilled specialists, called to perform dangerous tasks that involve not only the delicate and precise movement of heavy, expensive machinery, but the safety and well-being of the general public.

And if, like surgeons, some of them developed over-inflated egos, lone wolf habits, and a distrust of outsiders, that was to be expected.

In her teens, Nunez hero-worshipped the Tampa Bay pilots. "They walked tall," she said. "Even the ones that were not my dad's friends. When it came to a ship, they were kings. They were the master of that piece of metal. They would talk nothing but shipping when they were together at social events."

Because piloting is a more or less solitary business, many of the men did not have occasion to be in the same room at the same time unless they really desired the company. And the "three weeks on, three weeks off" rotation meant that Wiltshire, as manager, could keep the eighteen pilots divided into two distinct groups as he saw fit.

The two "sides" were unavoidably brought together when Florida created the Board of Pilot Commissioners. Now, instead of avoiding and being distrustful of one another, they had a common enemy.

"My father was against the change because he thought that they were losing control over the applicants," Nunez said, "and they would be given 'the hire' under the guise of diversity."

Harbor pilots saw the creation of the board as the ultimate act of governmental meddling into what had traditionally been a private matter. But the state was responding—slowly, as usual—to complaints about rampant nepotism and cronyism in the pilot ranks. Old buddies, and the old buddies' sons and cousins and nephews, were getting hired regardless of their qualifications. And there were no African American or female pilots in the state.

The enabling legislation specified who would make the big decisions:

> The board shall be composed of 10 members, to be appointed by the Governor, as follows: five members shall be licensed state pilots actively practicing their profession; two members shall be actively involved in a professional or business capacity in the maritime industry, marine shipping industry, or commercial passenger cruise industry; one member shall be a certified public accountant with at least 5 years of experience in financial management; and two members shall be citizens of the state. The latter three board members shall not be involved in, or have any financial interest in, the piloting profession, the maritime industry, the marine shipping industry, or the commercial passenger cruise industry.

The Board of Pilot Commissioners was also given the power to set pilotage rates, review and amend the number of pilots in each port, and take disciplinary action if warranted. For the pilots, Florida Statute 310.011 was Big Brother.

Wiltshire made it clear to his daughter that he neither liked nor trusted the Board of Pilot Commissioners. "They were not necessarily an advocacy group for the pilots," Nunez said. "They were more of a governing board, and they wanted to get rid of the riff-raff."

* * *

On October 18, 1976, John Lerro became the first pilot hired for Tampa Bay by the Board of Pilot Commissioners. He'd aced the Florida license exam and was issued a state license, and at the end of the month he officially resigned from the Panama Canal Company. In Tampa, he was designated Unit 19.

The family bought a small concrete-block house in Odessa, outside Tampa on the northern border of Hillsborough County. They decorated it with Tiffany lamps, smart paintings, and Oriental rugs. Lerro liked the place because it had a little pond in the back, which he came to call Lake Lerro; he bought an old rowboat and would spend hours rowing from one side to the other, timing himself.

For a while Sophie kept occupied by teaching a class in ballet at the

Fig. 12. Happier times: Lerro and his fellow pilot and best buddy Cyrus Epler goof around during renovations to Epler's cottage on Egmont Key. Lerro did not have many friends among the Tampa Bay pilots. Courtesy of Nancy Epler Calfee.

YMCA in Clearwater. Charles, now eleven, was enrolled in a nearby elementary school.

"My father loved all kinds of music, including bluegrass, blues, pop, folk and classical," Charles Lerro recalled. "We were fond of the Italian composers, but the only one we had recordings of was Vivaldi. We would hear others on the radio, but we never knew who they were."

One of Lerro's favorite albums was the 1975 *Suite for Flute and Jazz Piano*, a collaboration between pianist Claude Bolling and flautist Jean-Pierre Rampal. He bought Sophie an expensive flute made by Boston's William S. Haynes Company, which provided the instruments played by the great Rampal.

"And he collected all kinds of valuable art," Lerro's son recalled. "He had smuggled antiquities from Rome that were thousands of years old. He also had Picasso and Piranesi. I remember one piece he had was a Mayan face mask. I don't know where he got it, but I look forward to telling him about some of the Mayan-speaking friends I later made."

In 1978 Sophie became one of the merchant marine's first female radio operators and accepted a job with Texaco. This required her to be away from Tampa, away at sea, for those same lengthy stretches that her husband had just given up.

As was the rule, Lerro's first thirty days with the Tampa Bay Pilots Association were consumed with "observer" trips, up and down the bay, with each of the veteran pilots. Judy Nunez had started to work with her father, manning the office phones and the marine radio, and she quickly saw that Lerro—the new guy—was being sized up.

"The truth of the matter is, he had to prove his worthiness," she said. "And when he came in, *neither* side was receptive to his hire. My father was angry even before he met Lerro—Tampa is one of the shallowest channels, and not only had Lerro never navigated it, he had never even been *on* the channel."

Still, with his quick wit and his willingness to please, Lerro made a few friends. "He was a nice guy, and he had a good sense of humor," said Gary Maddox, who joined the association shortly after Lerro. "I would say a typical New Yorker, but in a good way, because I'm a New Yorker. He liked to talk, kinda like me."

Others took a dim view of the state's first hire and never warmed to him, even over time. "He might have been a fine ballet dancer," said pilot Robert Park, who'd been transiting Tampa Bay since the 1950s. "But he was not a pilot."

Park retired in 1986 after twenty-eight years on the bar in Tampa Bay. Interviewed in 2010, he still held strong opinions about John Lerro.

"He was the wrong guy in the wrong profession," Park said. "I had nothing against John. He was a nice kid. But he just shouldn't have been a pilot. He just didn't have what it takes to go aboard and take command of a ship and pilot it for fifty miles. He just wasn't able to do that. If he'd have kept going to sea, he probably would have got along fine."

6

Shoot for the Hole

Would the accident have happened if I hadn't gone along? Forget about whether you've got two guys on board or not. What if you had gotten on the pilot boat two minutes later because somebody had to go to the bathroom? Would you have gone up to full speed at the Sea Buoy two minutes later because you weren't making introductions? Two minutes was all that mattered.

Bruce Atkins

In the late afternoon of May 8, 1980, Lerro took a ship out of Port Manatee. He was just back from his three weeks off, and there was plenty of work waiting for him on the assignment board at the Egmont station. In the evening, he was originally scheduled to board MV *Summit Venture* and bring it to the phosphate docks in Tampa.

Agent Stephen Morrill, who represented International Ore and Fertilizer, had made the charter arrangements for *Summit Venture*. But Morrill had a party to attend that Thursday night, and so he changed the plan: the bulker would instead move from its anchorage at the Sea Buoy, the deep-water "holding area" where inbound ships awaited their orders six miles off Egmont, in the predawn hours of Friday, May 9.

"Because the ship would be going directly to the loading berth and because the Rockport terminal could load phosphate rock into the ship faster than ballast-tank pumps could pump out the ballast water, I told the captain to go ahead and pump out his ballast water at that time," Morrill said. "He would come in bow-high, but it would only be transiting Tampa Bay, and that would not be a problem; we did this all the time. I went off to my party and had a good time."

On the ship, Captain Liu Hsuing Chu ordered the ballast tanks emptied. Located at several strategic points inside the hull, these large tanks were pumped full of seawater to weigh the vessel down, so that it would ride lower and more heavily in the water—making for a smoother passage through potentially rough seas. Inside shallow Tampa Bay, however, a heavy ship could easily ground should it stray from the narrow dredged channel. And anyway, a lighter ship moved faster—time being money, after all.

So Lerro was to board *Summit Venture* at 5:00 the next morning. He retired to the deputy pilots' cabin after requesting a 4:00 a.m. wake-up call. The call was a few minutes late, and it was 4:25 before he walked out into the misty predawn and headed toward the "Big House," the cabin that served as reading and radio room, dining hall, and assignment center.

It was dark, and drizzly, and a lazy blanket of fog floated over the water; even so, the temperature was in the mid-70s, typical for springtime on the Gulf Coast of Florida. Peering through the wisps toward the northeast, Lerro could just make out the light on Buoy 11, two miles distant, which marked Mullet Key and the far edge of the shipping channel.

This was normal procedure for Lerro. If Buoy 11 could be clearly seen from the beach at Egmont, that meant at least two miles of visibility. There were many more items on his mental checklist, but eyeballing Buoy 11 was always the first.

Inside the Big House Lerro nodded hello to Bruce Atkins, who was to ride along as an observer that morning. May 9 was his thirtieth day as a trainee, and Atkins, who'd also been on an outbound ship the afternoon before, was restless. Unbeknownst to Lerro, he had called the association brass and asked to be relieved of this particular obligation.

"I would have preferred not to," Atkins recalled. "It was as much that I wanted to go on my own as 'Let's be done with this. Let's not make this trip.' But the rationalization was, I'm going to be here for the next forty years, and why stir the pot? You ride with twenty-nine, why not ride with the thirtieth? It's one ride up the bay, and that's it."

Because they had brought his tankers into Tampa Bay for years, Atkins was friendly with most of the local pilots. He had ridden with them all—Lerro included—and felt he knew each man's methodology well.

"You get a vast exposure to and a vast experience of pilots in all ports," Atkins said. "Some of them are excellent. Some of them you can see, in all situations, are unflappable. And in all situations the correct decisions seem to come kindly, and orders are issued with firm authority without getting all excited and cranked up.

"The real good ones tend to not get themselves in those pickles. They tend to anticipate them. Tampa had a lot of those pilots that were like that."

Atkins had seen Lerro in action and simply didn't expect to "learn" anything from him. "In piloting, you need to be thinking two or three decisions ahead of time," Atkins said. "Because what you do now on a 700-foot vessel is going to affect the next thing you do—and the thing after that, and the thing after that. What you get yourself into or out of, at a particular point in time, affects it down the road.

"Some pilots have that knack, and those are the pilots that I tried to emulate, and certainly enjoyed working with. I wouldn't necessarily suggest that John was in that class. John was in the next class, that would not, as a ship's master, give you the greatest level of comfort 100 percent of the time. Because he didn't exude that confidence, and so you always felt that."

Still, Atkins shook off his doubts, sipped his coffee, and waited for Lerro—his boss, like it or not, for the duration—to finish his own preparations. Lerro radioed *Summit Venture*, anchored at the Sea Buoy—and instructed Captain Liu to sit tight while he made inquiries about the weather.

In the Lloyd's Registry he looked up Motor Vessel *Summit Venture* and its vital statistics: a handysize bulk carrier, a.k.a. bulker, built in Nagasaki in 1976, owned by Hercules Carriers, a subsidiary of the Hong Kong firm Wah Kwong Shipping, and registered in Monrovia,

Liberia. The crew was Chinese. Operating under a "flag of convenience" was common practice: the shipowners registered the vessel in another country, and flew its ensign, in order to get around certain regulations governing hiring, training, and taxation. *Summit Venture,* four years old, 20,000 gross tons and 606 feet in total length, was a Liberian ship in name only. The single-screw vessel was equipped with a Sperry Mark IV course recorder, driven by a gyrocompass repeater motor.

He also listened to the scanner; on Channels 13 and 10, the Tampa pilots discussed ship and tug traffic and, most important, the weather. If the elements were dishing out anything particularly nasty, it would come up on the pilot channels. "I heard some conversation," Lerro said later, "but nothing about poor visibility."

Still, Lerro advised Captain Liu that he was going to delay boarding *Summit Venture* for a few minutes, while he continued with his checklist. He had a flood tide, moving in from the Gulf at approximately one knot. Earl Evans, piloting the bulk carrier *Good Sailor* outbound from the Port of Manatee, reported that he was just passing under the Sunshine Skyway Bridge; Jack Schiffmacher was pulling out of Tampa on the empty gasoline tanker *Pure Oil.*

Channel 16 was reserved for safety and distress as well as official Coast Guard business; had there been a warning issued about severe weather, the Coast Guard would broadcast it repeatedly on Channel 16 and refer the pilots to another frequency for additional information.

Aboard *Egmont,* the 50-foot pilot boat ferrying Lerro and Atkins to their assignment on *Summit Venture*, Lerro spoke with the pilot of the tugboat *Dixie Progress,* which was taking a barge inbound, not far away in Egmont Channel. Three miles' visibility was reported, which Lerro determined was fine for what he was preparing to do. "I figured I'd start the ship up, and if there [were] any problems we could anchor before we got anywhere," Lerro said.

In the forty-five minutes it took for *Egmont* to reach *Summit Venture,* Lerro listened in on the National Oceanic and Atmospheric Administration's "weather channel" on the pilot boat's VHF radio. No warnings had been issued for Tampa Bay.

At 6:20 a.m., in near-total darkness, the pilot boat came around the stern of the big ship. On the port side, out of the prevailing wind, a crewman dropped a rope ladder, and *Egmont* pulled up alongside. With both vessels in motion, Lerro and then Atkins grabbed for the ladder, swung themselves over, and began scaling the black-and-green steel body of the big bulker, like insects crawling up the flank of an elephant. They climbed near forty vertical feet of freeboard and swung themselves over the rail and onto the deck. The pilot boat returned to Egmont Key.

At that moment it wasn't raining at all. May 9 was starting to look pretty routine. Atkins, itching to get his deputy pilot certification, would finally be done with "observing." Lerro would be free to get to the bank in the afternoon and close on the loan that would buy him into the Tampa Bay Pilots Association.

The sun rose at 6:54. The first gulls and pelicans of the day had materialized and were making tentative dives for food in the breaking dawn.

* * *

Liu had anchored *Summit Venture* at the Sea Buoy in the late afternoon on Monday, four days earlier, after bringing her across the Gulf of Mexico from Houston, where the thirty-member crew had unloaded a cargo of steel from Japan. Most of them were veteran mariners, and many had served together on *Summit Venture* for eight consecutive months. The journey across the Gulf from Houston had taken four days, all of them uneventful.

When they clambered onto the deck of *Summit Venture*, Lerro and Atkins were greeted politely by Chief Officer Chan Chim Yee, who spoke in clear but broken English. When he asked the pilot his name, Lerro pointed to a label on his walkie-talkie—J. E. LERRO—and introduced Atkins.

The three men ascended the five flights of narrow stairs to the ship's wheelhouse, where the Americans were introduced to Captain Liu. The captain officially relinquished command of *Summit Venture* to Lerro, his compulsory pilot.

The wind was light, out of the southwest. Liu answered Lerro's questions about the engine speeds (full ahead: 11 knots; half ahead:

9 knots; slow ahead: 6 knots; dead slow: 5 knots) and about compass error (none appreciable); the current draft of the vessel (9 feet forward and 21 aft); and radar wave length (both three- and six-mile ranges, interchangeable and invaluable for plotting a course through spotty weather and alternating vessel traffic). Although there were two radar units in the wheelhouse, only the starboard one was being used. This, explained Liu, was because the two units, when operated simultaneously, tended to cause interference with each other.

Using a flashlight, the chief officer showed the pilots the location of the various dials and gauges that indicated wind speed, wind direction, ship's speed, angle of list, rudder angle, ship's rpm, and ship's time.

Satisfied, Lerro addressed the first order of business, which was to get the ship into Egmont Channel, the first leg of the journey. Lerro gave the conn to Atkins, who relayed the course to helmsman Wong Sau Gnok and put the speed at half ahead. West of Egmont Key, between Buoys 8 and 10, *Summit Venture* overtook *Good Sailor.*

When they came abeam of the Egmont Key lighthouse, Lerro glanced at the ship's clock. It was 7:06 a.m. When a vessel passed the lighthouse, with Buoy 10 abeam, it officially entered Tampa Bay. According to procedure, the portside tugboats and dock personnel in Tampa were notified at this marker to make ready for the ship's arrival.

As Atkins increased speed to full ahead, and adjusted the course to bring the ship into the next stretch of Mullet Key Channel, Lerro lowered his binoculars and advised Captain Liu to post a lookout and an anchor watch on the bow. Abeam of Buoy 11, he'd seen the first hint of rain. "We have rain squalls all around," he said out loud, to no one in particular.

Liu made a phone call, and the orders were given. Chief Officer Chan was watching for each marker through binoculars, and as they came up he wrote their numbers, and the time they passed, in the ship's log.

Soon it began to rain harder—more than a drizzle, Lerro thought to himself, but not bad enough to obscure the picture on the operational starboard-side radar screen. They were in the main shipping channel now, on a direct course for the Sunshine Skyway Bridge. The long concrete and steel structure was clearly visible as a flat yellow line on the radar picture.

At Buoy 14, 2.3 miles from the bridge, Lerro spoke with Schiffmacher aboard the *Pure Oil,* simply to establish contact. The light tanker was 3 miles to the west of the Skyway. Once they were both closer, they would reconvene on the radio and make plans for the eventual passing in the channel.

"Let me take the conn," Lerro said to his co-pilot. "This is squally weather. It's intermittent. It's not a pleasant day. Let me take the conn until we get past the bridge, and let's see what the weather does on the way." He instructed Atkins to monitor the radar, which was showing no rain clutter, and reduced speed to half ahead.

They were closing in on the Combat Zone.

The final buoys in the channel, before the necessary dogleg turn into Cut A and under the bridge, were 16, on the starboard side, and 15, to port. These came and went at 7:23, and that was Lerro's signal to begin looking for 1A and 2A, which marked the awkward course change through the 800-foot gap under the highest point of the Skyway.

He was unable to make visual contact.

Atkins, his eyes fixed on the radar screen, spotted 1A and 2A. Just two-tenths of a mile dead ahead, he told Lerro.

As *Summit Venture* chugged past Buoy 16, the sky exploded. Day became night. "Torrential rain hit the vessel, and it was just solid water in front of us," Lerro would testify.

Atkins's radar picture failed.

Captain Liu said nothing.

Lerro ordered the anchors made ready. Chan hurriedly left the wheelhouse, put on his rain gear, and proceeded forward to assist the bosun and the carpenter with the anchors.

Lerro thought fast. Whether he was ready or not, it was Skyway time. Going hard to port would be foolhardy, because of the unknown location of *Pure Oil.*

If he could get both hooks down, he might be able to turn slightly to starboard, and anchor just outside the channel.

“A hard turn to starboard was out,” Lerro testified, “because I’d present a broadside to that wind, and that wind was too much for that ship to take broadside. That was a light ship and that would surely affect it more than the wind from aft.

“I was thinking about taking the ship hard right, but I didn’t because also at the same moment, almost or just after I thought about doing it, the lookout said, and I know the exact words he said—or the exact words as I got them. He said ‘Buoy starboard bow.’

“That may have been the captain relaying it to me. It may have been somebody else. But I heard somebody say ‘Buoy starboard bow.’”

In that same instant Atkins reported the momentary reappearance of the turn buoys on the radar. We’re in the channel, he told the pilot.

That was good enough for Lerro. “The best thing to do, now that I had a position, I had a buoy sighting, was to shoot for that eight-hundred-foot hole.”

Immediately, he changed speed to slow ahead, and ordered Atkins to come to the next course, 18 degrees to port, from 081 to 063 degrees. Atkins relayed the order to the helmsman, and *Summit Venture* began its final left turn, a wide swing around that under normal circumstances would enable the ship to sail safely under the highway bridge.

To counter the effect of the strong wind out of the southeast, they had kept to the extreme right side of the channel during the transit and would remain there during this turn. Even without visual or radar cues, should *Pure Oil* turn out to be passing underneath the Skyway at the same time—on its designated side—all would be well.

And then Lerro saw the Skyway.

And then 1,297 feet of bridge deck and superstructure fell.

It was 7:34 a.m., May 9, 1980.

7

Tough Old Bird

Wesley MacIntire had a hard head, and he was proud of it. He'd been in more life-threatening scrapes than he could remember, and every time, he always joked, his thick skull had come between him and the end of the line. That, and some sort of divine intervention. Or maybe it was dumb luck.

His friends had a nickname for him: Tough Old Bird.

All he knew was that he should have been dead a half dozen times, and something had miraculously spared him. On paper MacIntire was a Protestant, but he wasn't really a religious man; he tended to shrug things off, and his irreverent sense of humor was one of the things people liked best about him.

He'd left Springfield, his Massachusetts hometown, at the age of seventeen to follow his older brother into the United States Navy. He wrote long letters about his time served, his buddies, and his misadventures to Betty Broadbent, who'd been his sweetheart since they were both twelve years old.

From Oran, North Africa, where the Americans were attempting to uproot the Vichy French in 1942:

> As soon as we got inside the harbor we met a French destroyer that was in for repairs. It opened fire on us, and after about ½ hour of fighting, they hit our water line and our ship started to sink . . . we had no sooner gone over the side when the French ship hit our magazine—boy, did that ship blow up! I went under water and didn't think that I was going to come up. A lot of the fellows never made it but I did get ashore. We were taken prisoner and what a place, nothing to eat but bread and wine and I got so drunk that I didn't know what I was doing—I tried to escape out of the window. I dropped about 15 feet to the ground but didn't get far when I was caught and brought back. We stayed there for four days and then Army tanks came into town and hundreds of Army men, and we were set free on Nov. 11, at 8 in the morning. And then after a few days we got back with the Navy.

On D-Day, June 6, 1944, he was on a landing craft storming the French coastline at Normandy, alongside thousands of other terrified young soldiers. As they approached the beach a German shell tore through the vessel's engine room; MacIntire came up sputtering through bloody surf, broken bodies bumping him like warm and nauseatingly supple driftwood. The noise was deafening.

On instinct and adrenaline, MacIntire ripped a pistol from the uniform of a dead sailor and ran, firing blindly, to the safety of another American landing craft.

Back in Boston, after back surgery, he found out that he was the only survivor from his outfit. MacIntire suffered a nervous breakdown and was honorably discharged for medical reasons.

Wes and Betty married in 1945, just as soon as he got back from his tour of duty, and by 1950 they had two kids—Lawrence and Donna—and began making payments on a small house in Wilbraham, one of heavily populated Springfield's eastern suburbs.

There was a small lake behind the house. "After driving trucks all week long," Donna remembered, "he'd come home and soap himself down—he kept a bar of soap under the pier—and then he'd jump into the water and stay under forever. I never thought he'd come up."

He took a job driving cargo for Rich Products, hauling big rigs full of frozen dairy goods from Boston to Buffalo, New York. This led to other trucking jobs, and when he wasn't making long-distance runs he was inspecting trucks, or working on their engines, for his employers.

Wes had always been mechanically inclined. His dad had taught him how to repair car engines—as a teenager, he used to tear up Springfield roads in his rebuilt jalopies—and in the service, he had proudly served in his ship's engine room.

He drove trucks, fixed trucks, and inspected trucks for thirty-five years, and Death rode shotgun a couple of times on his interstate drives. On a highway outside Worcester, he was behind the wheel of an asphalt truck when a station wagon cut him off, sending driver, vehicle, and cargo into a jackknife. Wes MacIntire flew through the windshield into a railroad crossing.

He survived, he told his family, because he'd landed on his head.

In Batavia, New York, a sudden stop sent him flying, again, into the windshield. He was scalped, pink and bleeding, like an unlucky pioneer crossing the plains in an Old West movie—once again, however, his head had saved him.

Like so many New Englanders who'd had enough of the snow and ice, Wes's parents had retired to Florida in the '50s, and Wes, Betty, and the kids began making annual visits to their modest home in Gulfport, a small, friendly Pinellas County community along the bay rim of St. Petersburg.

Once Larry and Donna were grown and out of the house, Wes and Betty decided to pull up stakes and make the big move, too. Betty was often in poor health, and the Florida sunshine was good for her. In 1976 they bought a small white stucco house, not far from where Wes's now widowed mother lived in Gulfport.

Officially, Wes was "retired," although he really just wanted to stop driving for a living. Financially, retirement was never an option. In Florida he took whatever jobs he could find, training drivers, inspecting trucks, or helping with fleet maintenance. In the wintertime his friend Stuart would come down from Springfield and stay at the house in Gulfport. They began a part-time roofing business.

The year before the Skyway accident, Wes took a nasty fall off a neighbor's roof, grabbed at a protruding metal edge, and nearly severed a finger before landing, bloodied, in the shrubbery. Dr. Edgar Buren, a plastic surgeon, stitched him up at St. Anthony's Hospital. Soon the Tough Old Bird was back on the roof.

* * *

On Wednesday night, May 7, 1980, Greyhound's Chicago-to-Miami bus, no. 4508, was right on schedule, huffing out of the downtown terminal at 11:30 p.m. The twelve-hundred-mile journey would take nearly two full days. Only one passenger, thirty-seven-year-old truck driver Mel Russell, got on at the start and planned to remain on the coach for the long trip to Florida; he intended to disembark in Sarasota, where his estranged wife, now living with her sister, had agreed to take him back.

As the Greyhound made stops in major cities, fresh drivers took over to tackle each successive leg. Passengers came and went. In Montgomery, Alabama, five students from Tuskegee Institute boarded, all bound for Miami. It was Mother's Day weekend, and Sharon Dixon, Yvonne Johnson, John "Chip" Calloway, Duane Adderly, and his girlfriend Laverne Daniels—all close friends in their early twenties—had just finished the semester and were heading home for family visits.

In Tallahassee Wanda McGarrah found a seat and settled in with her six-month-old daughter, MaNesha. McGarragh's mother, who had a birthday just a few days after Mother's Day, lived in Ft. Lauderdale and was anxious to see the baby. Tawanna McClendon, who also got on the bus in Florida's capital city, was a student at Tallahassee Community College. Like the kids from Tuskegee, the twenty-year-old was celebrating the end of another semester by journeying home for Mother's Day. McClendon's mother was in Palmetto, just over the Sunshine Skyway Bridge near Bradenton.

An eleven-year veteran of Greyhound Bus Lines, Michael Curtin had as his regular shift the last leg of the Chicago-to-Miami run.

Curtin and his wife lived in Apollo Beach, on the other side of the bay in Hillsborough County, and it took him about half an hour to drive to the St. Petersburg terminal, where he would take the wheel of No. 4508.

On the morning of Friday, May 9, he watched the worsening weather through his car windows as he crossed Gandy Bridge, one of the two thoroughfares that connected Hillsborough and Pinellas counties, over the northernmost part of Tampa Bay.

During the forty-five-minute layover in Tampa, Duane Adderly had put in a quick call to his father in Miami, just to check in and report on the bus journey. "Daddy," he said, "I'll be home in a few hours."

With stops, the run from St. Petersburg to Miami would take seven hours. No. 4508 left St. Petersburg at 7:05 a.m. Among those getting on were six Canadian tourists bound for Miami—two couples from Newfoundland traveling together, the Hudsons and the Browns, and elderly Manitoba sisters Lillian Loucks and Ann Pondy. A third sister, Stella Trush, was supposed to make the Florida trip but had slipped on the stairs and injured her leg. She convinced her siblings to go on and enjoy themselves while she stayed home and convalesced.

Margurite Mathison, an eighty-two-year-old St. Petersburg widow, was going to take a cruise from Miami to the Bahamas; Gerda Hedquist, ninety-two, planned to fly back to her native Sweden to visit relatives. At the time there were no international flights out of Tampa, and the arthritic Hedquist had to get to Miami to board her plane. She chose to leave the driving to Greyhound.

Once his twenty-five passengers had settled in, Curtin pulled the Greyhound door closed, and the bus eased out into the early morning traffic, toward U.S. 19 and the Sunshine Skyway Bridge.

* * *

Wes MacIntire stepped out into a chilly dawn at a few minutes past 7:00. A steady drizzle had turned his front lawn—a broad expanse of millions of tiny white landscaping stones—into a jigsaw of puddles that spilled and widened with the spittle of rain, even as he stood watching.

Many retirees preferred this kind of yard, which required virtually no maintenance, except when the weeds grew up around the flat walkway stones, or one of the not-so-carefully tended shrubs gave up, dried up, turned brown, and died. At least there was no mowing to do. In the Florida sun, particularly in the brutally humid summer months, people had been known to keel over from heat stroke while pushing a lawn mower and lie sprawled on the freshly trimmed green grass until somebody found them there.

At fifty-six Wes was no spring chicken, but he still had the broad shoulders and strong hands of his Navy days. He had kept himself in reasonable shape and remained an avid swimmer. He liked to swim laps at Gulfport Beach, a stretch of bayside just a short walk from his house.

Significantly, he had never taken up cigarettes.

Walking swiftly across the muddy expanse of stones, he heard the low rumble of thunder in the distance. The skies over Gulfport were chalky gray-black, even as the sun attempted to vault the treeline and poke its first tenuous strands through the morning clouds. It was still drizzling, and his shoes threw back weak little waves as he strode hurriedly to his vehicle.

Wes MacIntire loved his little blue Ford Courier. The pickup truck was a 1974 model—already six years old—but it ran like a top. He called it his baby; since he'd given up driving the big rigs, MacIntire had owned a succession of large, gas-guzzling cars, but now the Courier, he told Betty, was all he needed. The low, comforting rumble of the truck engine was pleasantly familiar. He had bought it, used, in 1979.

He climbed into the cab, pulled the door closed, and shoved his key into the ignition. On the radio the morning news was full of talk about the funeral of Josip Broz Tito, the president of Yugoslavia, on Thursday, and the ongoing exodus of Cubans arriving on Floridian shores on makeshift rafts and any other thing that floated. The so-called Mariel Boatlift was becoming a national crisis.

President Jimmy Carter was still licking his wounds over a failed attempt at rescuing the fifty-two Americans held hostage by Iranian militants. And a U.S. Coast Guard Marine Board of Inquiry was knee-deep in its investigation of January's *Blackthorn* tragedy. Wes switched the radio over to the Christian music channel that Betty always had playing in the house.

For about eight weeks he had been making this drive into Manatee County, to the south across the Sunshine Skyway, every weekday morning. He'd accepted a job offer from Palmetto Meat Dispatch, Inc., which owned a small fleet of trucks. He taught the fine art of the long haul to novice drivers, worked on engines, and sometimes even drove a load of cargo himself. Before Meat Dispatch, he'd done the same for the Tropicana Juice Company, which maintained a distribution center in Bradenton, the biggest city in Manatee County.

The rain began to fall harder as he edged his blue baby out into traffic on busy U.S. 19, the only road that directly connected Pinellas County, where the MacIntires lived, and the cities to the south.

Fifteen minutes after he'd pulled out of his driveway in Gulfport, he approached the toll booth that would direct him onto the Sunshine Skyway Bridge. The Courier's windshield wipers were slapping loudly.

As MacIntire slowed the truck to toss his quarters into the webbed tollbooth basket, he noticed several cars crawling along in the right lane, their emergency lights flashing, as if waving the white flag of surrender to the rain, which had begun coming down in sheets and was being blown so hard that it whipped horizontally across their hoods. He saw vehicles pulled over in the Skyway rest area, their drivers unwilling to brave the long drive over open water in such awful conditions.

It was not hurricane season, though; the big blows generally didn't start cooking until August or September. As bad as it was out there, Wes knew it was just the way springtime weather worked in Florida. In ten minutes, the sun would probably be shining.

For a minute or two he did consider pulling off the road. As a professional driver, Wesley MacIntire was only too aware that driving in heavy rain could be hazardous; he'd had plenty of close calls while hauling canned goods, or cases of disposable diapers, or frozen vegetables up and down hilly New England highways. But he was intimately familiar with this route and was not easily scared.

So he continued, and once the four-lane approach road separated into twin causeways, two lanes going south, two going north, he felt better. He was headed south across the bay, and although traffic was creeping along cautiously, at least everybody was going in the same direction.

Now there wasn't any possibility that a disoriented driver headed the other way, coming toward him, could panic in the downpour, cross over into his lane, and smack into him head-on. In the old days, when there was just a single two-lane Skyway with no divider, that sort of thing used to happen a lot during wet gusts and heavy squalls.

As he began the part of his journey that would take him over open water, he relaxed a little in his seat. He'd never liked seatbelts and hadn't bothered to fasten his. Almost immediately, too, the rain slackened, and MacIntire sped up a little, pulling ahead of other cars as they all approached the steep climb of the Skyway's main span. He pulled around a Greyhound bus, which was throwing off a lot of spray.

MacIntire enjoyed watching the huge cargo ships passing under the Skyway on their way into or out of the Port of Tampa, and sometimes he adjusted the Courier's speed so as to reach the bridge's highest point just as a behemoth was gliding underneath.

He shifted into third gear to begin the climb.

The wind seemed to come from all directions at once, and the Courier was pelted with intense, gusty blasts of rain. It wasn't pitch black, but visibility was limited to a few feet in front of the Courier's windshield—the tail lights of the vehicle in front, and even those only for a second or two until the wiper blades made another pass. The truck's windows were rolled up tight.

Just as it crossed the stretch of metal grating at the very pinnacle of the Skyway, before the long slope down again, the Courier began to pitch violently back and forth. Wes felt it bounce; his first thought was that the ferocious wind was now blowing up through the gaps in the grating. The road under him was wobbling.

He slammed down hard on the brake pedal just as the little truck slid into the left-lane guardrail, but the tires couldn't grip the slick metal roadway. "Oh God!" he screamed.

He was falling, but the road under him was falling faster.

The next thing he remembered seeing were the words SUMMIT VENTURE, painted white on a background of black, getting closer and closer, and just like that, Wes MacIntire knew where he was. The grey sky, the green water, the heaving dashboard and the bobbing asphalt blurred together.

This is it, he thought. *I've had it this time*. Even his hard head couldn't save him from this, he thought; Betty would get a kick out of that one. He braced for the impact he knew was coming.

8

The Abyss

> It's a horrible thing. Oh man, this is just awful. I wish the media would jump all over this thing . . . a boat going through a storm like this is stupid.
>
> **Dick Hornbuckle**

At 7:00 a.m. Eddie Bartels clocked out of work at the Seaboard Coastline Railroad in St. Petersburg and headed for his Bradenton home. The thirty-six-year-old car inspector had been making the graveyard-shift commute southward over the Sunshine Skyway nearly every weekday since 1972. He was on the road at precisely 7:02.

A cautious driver who boasted to his wife and friends that he'd never had a speeding ticket in his life, Bartels kept his GMC Pickup at a steady 25 mph as he navigated south along U.S. 19 on the straight shot to the Skyway. The rain was moderate, in his view, nothing to get concerned about. Still, it was a black morning, and he took his time.

Just after tossing 50 cents into the basket at the Skyway tollbooth, however, Bartels began to feel the wind wake up, blowing in consistently muscular gusts from the west—the right side of his truck. The drizzle turned to steady rain, and then to horizontally sheeting rain, punctuated—not at all unusual for Tampa Bay, where the weather was famously erratic—by short periods of eerie, pregnant stillness.

Then it would start all over again.

As he moved along the two miles of open-water bridge before the big climb, Bartels noticed water shooting up into the air through the drainage ports in the roadbed, little geysers like fire hydrants cracked opened by some unseen force.

At its worst, visibility was no more than two or three car lengths, but Bartels had braved this sort of storm before, and as he reached the top of the bridge, he felt the wind ease up—the steel superstructure, he knew from past experience, blocked out the wind for a couple of seconds, like that moment when the rain seems to die as you drive beneath an overpass. He was driving cautiously in the right-hand lane.

Quickly he scanned the road ahead for the tail lights of any vehicles he might have to contend with on the downward slope and the remainder of the open-water causeway to Bradenton. He made a mental note of a ten-wheel flatbed truck loaded with concrete blocks, driving several hundred yards in front of him.

The rain came at him again, noticeably weaker now, and as Bartels began his careful southward descent, he looked westward over the water, as he always did. And he saw it: *Summit Venture*, perhaps 600 feet—a single ship's length—from the roadway. It was in a wide turn, Bartels could see, pointed bow-first in a northeasterly direction.

Moving in the direction of the Skyway's support piers.

"I realized it was on a collision course with the bridge," Bartels told investigators. "To the point where I speeded up to an ungodly speed. Which I don't do."

In seconds he had cleared the descent, racing through the pulsating rain. "I knew the ship was going to hit," he said, "but it didn't register that the vibration and the shock I felt was actually the bridge being hit. Until maybe five seconds later, when she started rumbling like a long thunder roll." Then he heard the shrill, pained sound of screeching metal.

Bartels didn't look back. The last thing he remembered was speeding past the ten-wheeler and its load of blocks on the flat causeway, both vehicles safely on their way to the other side.

* * *

Donald Albritton was crawling along the southbound causeway in his 1977 Chevy Nova, gripping his steering wheel tightly, the windshield wipers pumping feverishly and the defroster turned up to maximum blast. Even so, he could barely see the nose on his face. The wind was blowing hard from the west to the east, and he had to stay focused on keeping his little car steady in the nasty weather and the upcoming ascent.

An ad salesman for the *Bradenton Herald* newspaper, Albritton had a fairly short commute from his home near U.S. 19 in St. Petersburg, about twenty minutes from the Skyway.

He began the slow rise up the main span, noticing, as he allowed himself a quick glance upward, the tail lights of another vehicle—a small pickup truck—at the very crest of the bridge, swing unnaturally sharply from the right lane to the left.

Visibility improved somewhat, enough for Albritton to glimpse the flashing red lights bolted into the upper edge of the superstructure, a warning for low-flying airplanes. He knew them well.

Silently, as if on a drive-in movie screen with the sound turned down, the red lights dropped and fell from view. They were just . . . gone. "I could see the lights cascading down like Roman candles on the Fourth of July," he said. "The bridge started violently shaking. It was almost like an earthquake." He didn't hear a thing.

Albritton opened his eyes wide and gasped; instinctively, he pumped his brakes and slowed to a crawl.

He was in the right-hand lane, just two or three car lengths from the spot where the Skyway's skeletal steel superstructure was riveted into the truss and concrete. He looked up again, and as he did he could swear he saw part of the steel girder pitch, wave and roll, and then drop abruptly from sight.

Again, there was no corresponding sound. Just the steady spray of the rain against the side of his car. He took a quick, deep breath—a terrified breath—and slammed the brake pedal down.

Albritton put the Nova in reverse and began slowly backing down in the right lane, the way he'd come. Halfway down, he realized several cars were approaching, beginning the climb. Still not sure what had happened, but operating on dread and adrenaline, he rolled down his window and waved his left arm wildly, hoping to get the attention of the approaching drivers. He pounded the side of the car and screamed.

Three or four cars passed him, driving in the left lane. Then a Greyhound bus. He couldn't tell if any of them had heeded his hysterical warning; the bus driver didn't look his way, or blow the horn. Albritton didn't see any brake lights. Soon all the vehicles disappeared from sight, over the crest of the Skyway.

Ken James had noticed the bus fly past just moments before. He'd been concerned about the brutal rain and hit-and miss visibility after driving away from the toll booth, and had briefly considered pulling to the side, joining the little group of cautious Friday morning motorists who'd already done so to wait out the weather.

As he passed the rest area—one of the last places he could have sought refuge before the shoulders fell away and the road turned to elevated over-water causeway—he noticed a small truck about 50 feet in front of him, with three palm trees standing in the back, their green fronds whipping in the cruel wind.

A flashing blue light appeared from somewhere behind his car. The state trooper passed him, moving fast, and James watched as the truck driver with the palm trees was pulled over to the side. The officer got out in the driving rain to talk to the man.

James was doing about 40 mph, in the right lane. He looked over just as the Greyhound went by in the left lane, traveling at what James later estimated was about fifty, followed almost immediately by a small, dark automobile.

As James crossed the flat causeway, the rain picked up again. The bus and the small car were no longer visible, and the two or three additional cars that blew past him—he'd reduced his speed to 25—vanished into the weather after only a few seconds.

He began to ascend the main span, and visibility improved once more. As he neared the apex, a motionless vehicle materialized in the right lane. Its flashers were on, and as James approached, he saw the driver—a man—waving his arm through the open window and slapping the side of his car.

* * *

Paul G. Hornbuckle had seen the truck transporting the palm trees. He didn't see the state trooper and his blue lights, but he had noticed that the truck driver was backing his vehicle up onto the roadside, so that the bed—and the trees—were facing into the pummeling wind.

That seemed like a wise choice, Hornbuckle thought; rather than risk the crossing, where the wind could toss his blowing cargo into the bay like a candy wrapper, the palm tree man had apparently decided to wait out the worst of it.

Hornbuckle had been known as "Dick" since his boyhood days in Georgia, although he never told anybody why. He attended Florida Bible College in the early 1940s—Billy Graham was in his graduating class—and over the decades had held down a wide variety of jobs. Somehow, he'd wound up in the used car business.

At sixty he was something of a mysterious figure, even to his friends. He'd been married several times, but he didn't like talking about it. What Dick Hornbuckle liked to talk about was business. He was a deal maker. A longtime St. Petersburg wholesaler and dealer, he crossed the Skyway several times every month, traveling south to pick up vehicles he'd purchased from other dealers. Hornbuckle had an office at Apollo Auto Sales on U.S. 19, less than a five-minute drive from his bachelor apartment.

On May 9, the quick-thinking, fast-talking car dealer was headed south, behind the wheel of his squash-yellow 1976 Buick Skylark sedan. Across the grill he'd attached a sheet of window screen, a common practice in Florida during the spring, when the air was thick with mating lovebugs. Their squashed carcasses, in the thousands, could play hell with a car radiator.

Hornbuckle had brought his pal Tony Gattus with him. Because he'd had a heart attack, the sixty-two-year-old Gattus had been forced to retire from his job with the Pinellas County Sherriff's Office—his duties had included searching incoming mail for contraband—and he was bored. Two or three times a week he went to Derby Lane, on the other end of St. Petersburg, to bet on the greyhound races. The guys at the betting windows knew Gattus by name.

Sometimes he spent the day driving cars across the state for Dick Hornbuckle.

Gattus had made this particular journey with Hornbuckle many times before. On May 9 they were on their way to dealerships in Sebring and Avon Park, about three hours south, to take possession of three new acquisitions; their cronies Jim Crispin and Kenneth Holmes had been recruited to drive the other vehicles back to the Apollo lot in St. Petersburg.

The radio was off, and conversation among the four men was kept to a minimum. Visibility was bad, then it would clear up, and then the sky would turn black again. Hornbuckle gripped the steering wheel hard and squinted through his thick eyeglasses. A small blue pickup truck passed him, followed by another car, and another, and a passenger bus. They were quickly lost to view in the enveloping blackness. He slowed to 25 and switched on his flashers.

Crispin, sitting in the passenger seat next to Hornbuckle, loudly suggested they pull over and wait for the worst to pass. But Hornbuckle had appointments to keep. It was, he would later testify, "as dark as night" as he started the climb up the main span. He could feel the big Buick sway with the force of the wind.

Near the top of the climb Hornbuckle came upon a vehicle stopped in the right lane. As he slowed, he saw Don Albritton waving his arm and pounding the side of his car. *I guess he's stalled*, Hornbuckle thought. *Or scared*. He moved into the left lane to ease around the little Nova, and reached the metal grid section at the top of the bridge. He began to hear that familiar baritone hum, deep and ominous like something chanted from beyond the grave.

Hornbuckle was just about to speed up when it hit him—instead of the other side of the bridge, the way down, there was nothing. It was as if someone had lowered a dark curtain halfway across the Skyway. He slammed on his brakes and the Buick's tires caught traction on the slippery surface. All four men dared to look out through the windshield. Before them, through the rain, which was subsiding again, there was *nothing*.

The Buick slid across the white center line came to a stop less than 14 inches from the abyss.

It was balanced at the tip of a 35-foot section of thin metal, dangling downward into space at a 13-degree angle, the steel teeth raw and exposed where the section's mate had been violently ripped away.

Hornbuckle, Gattus, Crispin, and Holmes threw open the Buick's doors in unison, fell to their knees, and carefully crawled up the slippery metal, back to the concrete roadway.

They had no desire to look back.

Gattus recalled: "I'm hollering to the others—the cars ain't going to see us or hear us. You can't see your hand in front of you. We're going to get run down."

Gattus, Crispin, and Holmes began to wave their arms frantically at the automobiles moving toward them: "The bridge is out! Stop!"

Hornbuckle, however, had taken a quick glance over his shoulder and realized that all four of the Buick's doors were still open; tenderly, he edged back down the broken grating, took the keys out of the ignition and walked around the car, carefully closing all four doors lest the wind take his yellow beauty another foot or two.

Hornbuckle came around to the trunk, keys in his hand, intending to extract his beloved golf clubs. He had a game the next morning. His three friends watched, incredulous. They shouted at him—"What are you, crazy? Dick, are you trying to get yourself killed?" Hornbuckle left the Buick, and the golf clubs, and as he rejoined his friends he began to consider their good fortune. Their incredible good fortune.

"I knew some insurance company would take care of my car," he later explained, "but those clubs were like an old pair of shoes. You get used to them."

* * *

Ken James figured there was probably an accident, a fender-bender, somewhere near the crown of the bridge. In this weather, such a thing wasn't unthinkable. He assumed the Nova's flashers were on because the driver was directing him around it. James pulled slowly into the left lane and began to pass the stationary car with frantic Don Albritton inside.

As he eased slowly back into the right lane and sped up, James caught sight of a yellow car—just its roof was visible as he inched forward—ahead of him. It was clearly stopped on the Skyway's metal

Fig. 13. The scene of the crime: with Dick Hornbuckle's Buick just inches from oblivion, *Summit Venture* is held fast by its port anchor chain, which was pinned to the bottom by tons of debris from the fallen Skyway. Courtesy of Florida Archives.

grating, the bridge's highest point. That was enough. He braked, hard, and turned off the engine.

"I saw four men coming toward me from the yellow car," James recalled. "The men were waving their arms and yelling to me that 'The bridge is out.'"

Over the wail of the wind, 150 feet over Tampa Bay, the hysterical men gestured with their arms and shouted at James to block oncoming traffic with his vehicle. He cranked the engine and backed it up sideways, covering a section of both southbound lanes.

James took a second to gather his courage; he got out of the car, craned his neck and peered over the edge—as far as he dared. There

Fig. 14. A matter of perspective: the same scene from water level, with the Hornbuckle car teetering 150 feet up at upper right. Courtesy of *Tampa Bay Times*.

was the massive *Summit Venture*, its bow nosing between two piers of the Skyway's northbound span. The roadway in front of him was gone. Giant sections of the steel bridge superstructure were sticking out of the water.

James noticed a body at the surface; he watched it slowly float away, in the general direction of Tampa. He forced himself to turn away and shuddered.

Another car appeared, a gold Pontiac convertible, creeping toward them; James ran to it and screamed at the driver that the bridge was out; he should turn around and drive back toward the north—there was a highway patrolman down below, pulled over next to a small truck with palm trees in the back.

Richard Baserap, on his way to his job as a welding instructor at a Sarasota vocational college, stopped his vehicle and attempted to process what James was saying. A driver on the northbound Skyway had stopped his car at the summit and was shouting a warning over the wind. "You'd better get out of there," the man yelled. "That whole thing is going to come down!"

Baserap grasped the situation and jumped back into the Pontiac. He grabbed for his CB radio, tuned to Channel 9 and shrieked that the bridge had fallen. He too had noticed the trooper's car, and the truck with the palm trees, and hoped to contact the officer down at the rest area. But Baserap got no response. So he switched to Channel 19, and immediately heard the excited chatter of truckers—and they were talking about the Skyway Bridge. Clearly, the message had gone out. He turned the Pontiac around and began to drive, as quickly as he dared, back down the Skyway incline, heading north. But there was no traffic; Baserap watched as the highway patrol sedan raced past him, blue lights spinning and siren shrieking, straight for the top of the bridge.

* * *

Whether by the catching of the anchor, or the full astern order, or the impact with the bridge, *Summit Venture* was stopped. Dead. The flare of her bow had clipped pier 2S about 54 feet off the waterline, above the puny "crash wall" built into the pier's base to deflect bumps from small craft.

Both spindly columns of pier 2S were completely gone, as was everything above them. The exposed steel rods, broken and twisted and freed after a decade inside a cocoon of concrete, beckoned like bony fingers from the grave.

Summit Venture's bow had come within several yards of the older span's supports; had Lerro not given the commands that he did, when he did, the ship likely would have kept moving and taken both bridges down.

LERRO: *Mayday, mayday, mayday, Coast Guard! Mayday, mayday, mayday, Coast Guard!*

COAST GUARD: Vessel calling Mayday, vessel in distress. This is United States Coast Guard, St. Petersburg, Florida. Request your position, nature of distress, and number of persons on board. Over.

LERRO: *Get all the emergency equipment out to the Skyway Bridge. Vessel just hit the Skyway Bridge. The Skyway Bridge is down! Get all emergency equipment out to the Skyway Bridge. The Skyway Bridge is down! This is a Mayday. Emergency situation. Stop all the traffic on that Skyway Bridge!*

COAST GUARD: This is Coast Guard St. Petersburg, roger. What size is the vessel that hit the bridge, over?

LERRO: *Stop the traffic on the Skyway Bridge! There are some people in the water. Get emergency equipment out to the Skyway Bridge now!*

COAST GUARD: This is Coast Guard St. Petersburg, roger. What vessel are you on, over.

LERRO: *Summit Venture, Summit Venture.*

COAST GUARD: Summit Venture, Coast Guard St. Petersburg, roger. What is the size of your vessel, and can you assist, over?

LERRO: *Cannot assist, we're 606 feet long, light draft. We cannot assist. We are in on an abutment. Stop all the traffic on the bridge. Send some vessels over here to render assistance. People are in the water.*

COAST GUARD: This is Coast Guard St. Petersburg, roger.

* * *

"That," recalled Bruce Atkins, "was very frustrating; that typical bureaucracy in the Coast Guard. You're calling a mayday, you're saying you struck the bridge, there's cars going off the bridge—stop the traffic at the tollbooths! That's the first thing. You've got people that are there at the tollbooths. Flick a switch! Throw up the red lights—nobody gets through!

"And the Coast Guard's going, 'Well, how big is your vessel? How many people on board?' *Stop the goddam traffic!* Stop the traffic and we'll sort out who we are later on! We'll answer all of your questions."

It was as if the ship's clock had stopped, lifeless and frozen over, while the world outside their window continued to spin out of control.

Atkins and Lerro had watched helplessly as one car after another flew off the roadway and nosed downward, each set of headlights illuminating the steady drizzle of rain. With each silent splash, they felt their stomachs tighten. "I just saw the cars come down, almost rippling in slow motion," Lerro would testify. "Tumbling off the bridge, one by one. No matter how hard I yelled, no matter how hard I struggled, no matter whom I tried to get to stop. No one . . . nothing . . . would stop.

"And they kept coming . . . they kept coming over the tear in the bridge and falling like toys in slow motion."

The pilots braced themselves as yet another set of headlights appeared at the apex of the broken bridge. They watched in horror as the wide yellow vehicle began to skid on the steel grating—and then came to a stop just a few inches shy of the edge.

As the doors flew open, and the car's four occupants emerged and began crawling up the slippery grating to the sure footing of concrete, Atkins and Lerro held their breath. Would the remaining gridwork give way? Would these four people plunge to their deaths?

They saw the men stand up and sprint the last few yards to safety. Clearly, word had been passed that the bridge was gone, and no more vehicles would make that sickening plunge.

The dead silence was broken by the crackle of the radio. The Coast Guard dispatcher appeared to have realized—at last—the enormity of the situation and instructed Lerro to stay on 16, the emergency channel.

"Pass all traffic on 16, break, break, pan-pan, pan-pan pan-pan," came the voice, now noticeably more urgent, using the universal maritime code for *breakdown*. "Hello all stations, this is United States Coast Guard St. Petersburg, Florida. Silence is imposed on this frequency. Seelonce Mayday. Seelonce Mayday. Silence is imposed on this frequency. This is United States Coast Guard, St. Petersburg, Florida, out."

The tugboat *Dixie Progress* was the first to respond. Captain Evans said he was in the area and would proceed to the bridge to render assistance. The Coast Guard operator informed the captain that people were in the water, then repeated the call for any vessels in the vicinity to head to the bridge to help, adding the information about the size of *Summit Venture.*

Four minutes had elapsed since the collision.

When the dispatcher called again, Lerro's voice was noticeably less frantic. His responses were short, his tone numb. Shock and helplessness were already turning to resignation as the reality of his situation began to sink in.

COAST GUARD: Vessel that hit the Skyway, Summit Venture, this is Coast Guard St. Petersburg, over.

LERRO: *Summit Venture back.*

COAST GUARD: Summit Venture, Coast Guard St. Petersburg, roger. What is your cargo, over?

LERRO: *We're light in ballast.*

COAST GUARD: This is Coast Guard St. Petersburg, roger. Where did the people come from that are in the water, over.

LERRO: *From the bridge.*

COAST GUARD: This is Coast Guard St. Petersburg, roger. Stand by 16, do not shift to any other frequencies.

LERRO: *Standing by.*

COAST GUARD: Summit Venture, Coast Guard St. Petersburg. Did either span of the Skyway go down into the water, over?

LERRO: *The west span went down, the west span.*

COAST GUARD: Roger, the whole west span, over?

(pause)

LERRO: *The west span is down in the water.*

9

The Only Fool Who Went in the Water

> Nobody was ever reviewed. He should have been. Everybody knew what he was. Kind of pitiful, really. But he just kept going, and going, and going, and he kept getting worse, not better. As you progress up the line, you handle bigger and deeper ships. God knows what would've happened if he'd got up to full pilot.
>
> **Robert Park**

Wesley MacIntire woke up suddenly, as if jolted from a nightmare, with a pounding headache, blood running down his face and a terrifying sense of claustrophobia. The air was close. He was cold and soaking wet. It took him a few seconds to remember that he'd just been looking straight into the unforgiving steel hull of a massive ship. Had he dreamed it? Where the hell was he?

Then he noticed the green water pouring in through the space where his windshield had been, and bubbles rising in steady streams from the battered hood of his little truck, which seemed to be parked at a curious angle. The headlights were still on.

He was at the bottom of the shipping channel, the truck's rear end planted in the silt 40 feet beneath the surface of Tampa Bay.

His little truck's fall had been broken by the port bow of *Summit Venture*.

Frantically, MacIntire forced open the top part of the damaged driver's side door until he'd made enough room to wiggle through. The water stung and the escape route was tight.

His Navy survival training, from forty years before, came rushing back.

Filling his lungs one last time, MacIntire pushed his 220-pound frame out of the truck and kicked for the surface. He could see a vague light, but he didn't really know if he could make it all the way through the blackness.

By the time he broke through—"like a bullet coming out of the water," he later described it—he'd swallowed a stomachful of seawater, and the first thing he did was vomit. Blood was streaming down his face from the gash on his forehead. His legs ached.

He was treading water under the intact north-bound span of the Skyway, sick and scared and gasping for weak, wet breath. A few feet away, just sticking out of the swirling green water, he saw a large segment of silver steel superstructure, so big that one end was on the bay bottom, the other poking through the waves. He managed to grab it and hold himself steady.

Through the cold drizzle, MacIntire managed to focus on the broken bridge a thousand feet away and 150 feet up. He fixated on a pair of bright headlights—Dick Hornbuckle's Buick—right at the dangling edge.

So that's what happened.

He tried to mute the sound of his pounding heart to listen for splashes. Or screams. He was, after all, a Navy swimmer, and surely he could help anyone else who might have fallen.

But there were no sounds. *I'm the only fool who went in the water,* he thought.

Then he heard a sickening mechanical groan from behind him, and the silver piece he held onto began to move. Swiveling his head, MacIntire beheld the massive black-and-green hull of *Summit Venture*, not 20 feet away; the ship was moving slightly, trying in vain to respond to the pilot's last orders: full astern. It swayed almost imperceptibly beneath the shattered bridge, straining against the anchor chain pinned by wreckage. A section of intact asphalt roadway and steel

Fig. 15. Following the collision, *Summit Venture* remained in place despite its most recent engine orders, full astern. The port anchor had been dropped and was pinned by tons of debris. Pilot Robert Park was summoned to extricate the ship from the wreckage and anchor it safely to the west. Courtesy of *Tampa Bay Times.*

guardrail, still connected, was strewn horizontally across the gargantuan bow, broken edges trailing over the side like beaten, defeated limbs.

I'm hanging on and I'm throwing up, MacIntire thought, *and now this ship is going to run me over.*

"Jesus, someone help me," he managed to yell, and almost instantaneously *Summit Venture's* carpenter Lok Lin Ming appeared on the deck, peering down at the tiny, bleeding man in the choppy water 40 feet below.

Lok shouted something, left for an instant, and returned. He threw down a long rope with a loop in it, which MacIntire put around under his arms. Then the carpenter and bosun Sit tossed a second lifeline: the very same pilot ladder that Lerro and Atkins had used for their climb aboard the ship a little over an hour earlier.

It took several attempts to get him to hold tightly; MacIntire had little strength in his legs. Finally, desperately, he swung himself back and forth in the froth until the ropes tangled around his body. The crewmen grunted and hauled him out of the churning bay and over the side, a bluefin tuna worn down, glass-eyed and inert, at the end of a longline.

It was 7:55 a.m. He'd been in the water for twenty minutes.

The lookouts carried MacIntire's limp, bleeding body down the endless length of deck toward the stern. Blood streamed down his pale face, turning his white moustache red. His eyeglasses were long gone. After assuring his rescuers that he was essentially undamaged, he was delivered to Sick Bay, where he would remain for the better part of two hours. His wound was cleaned, and a primitive dressing was applied.

His rescue encouraged the remaining crewmen, who had received the pilot's orders to line the deck and search for survivors. But as the gray waves licked the Skyway supports in a steady rhythm, the channel water would give up no more secrets. The sailors never saw another person.

* * *

Tampa Bay Pilots Association manager B. F. Wiltshire was on his three weeks off and attending a convention in Texas, so the dispatcher's frantic call went through to his emergency number—daughter Judy Nunez's house in Brandon, across the bay in Hillsborough County.

Nunez contacted her father, who began making phone calls from his hotel room in Dallas, where it was not yet 7:00 a.m. .

Because he lived on nearby St. Petersburg Beach, pilot Robert Park was the association member closest to the Skyway. Park, Wiltshire knew, was also on his twenty-one-day off-duty phase and was therefore likely to be at home.

It didn't matter that Park was perhaps Lerro's most vocal critic.

Within minutes of receiving the order, Park was on a pilot boat out of Mullet Key and bee-lining through the chop for *Summit Venture*. He was to stabilize the ship, assume control, and assess the damage. Park also knew Captain Liu, and had a nodding acquaintance with

Bruce Atkins. His priority, Park was told in no uncertain terms, was Lerro. "We knew what his personality was, and we didn't know what he would do or say," Park recalled. He thought the much younger pilot talked too much. Privately, he considered Lerro a complainer. A whiner.

"In a situation like that," Park said, "the last thing you need to be doing is talking, where somebody could ask you five years later 'Did you say this, did you say that?'"

Beyond taking charge of the ship, Park's assignment was to confer with Lerro, calm him down, and convince him to button up until the pilots' attorney arrived. And/or the Coast Guard investigators, who were already on their way.

Park was delivered to *Summit Venture* at 10:00 a.m. and ascended the gangway, the collapsible aluminum staircase the crew kept stored in the stern, behind the superstructure. The gangway, extended over the side to connect with pilot boats, was also used for nonpilot visitors—those who couldn't or shouldn't be expected to rappel up the freeboard hanging onto a length of rope.

The senior pilot raced up the five short flights of stairs to the wheelhouse and found Lerro standing motionless, silent, staring through the rear portholes, as if wishing he could will time to move in reverse. Park delivered his message, exchanged a few words of solemn solidarity with Liu and Atkins, then radioed for a pair of tugboats to maneuver themselves on either side of the ship and hold it in place, to keep it from knocking into the *other* Skyway span, to which its creeping bow had come perilously close.

Park went downstairs again and made his way down the long deck to the bow. Crewmen—some in rain gear, some in their regular blue work clothes—scurried back and forth. Park intended to mount the foc'sle and get a look at the two anchor windlasses.

He found the entire foc'sle head inaccessible, buried under fallen roadway and grotesquely bent steel. The yellow forward mast, location of the navigational light, forward deck lights, foghorn and whistle, had poked a clean hole through the road section as it landed on top of *Summit Venture*. The mast was sticking straight up through the asphalt.

Fig. 16. The undamaged 1954 span would be opened to two-way traffic two days after the collision. The operational Skyway would remain like this—with the broken span at eye level—until the replacement bridge opened in 1987. Courtesy of *Tampa Bay Times.*

The port anchor, Park discovered, had been let out one shot of chain, which was enough to rest it on the bottom but not enough to hold *Summit Venture* in place should the weather worsen again. It was jammed by fallen debris and wouldn't budge.

Park considered the situation. The most immediate threat, he reasoned, was to the undamaged eastern Skyway span. Even with steadying tugs positioned on either side of her, *Summit Venture* was still moving with the waves and currents. Conceivably, another couple of strong gusts could blow her hard enough to bring the other bridge down.

He therefore issued an order to push the silent freighter away to the west, as far as she would go with the port anchor still out and stuck, for the time being, on the bottom. This turned out to be about a ship's length from the bridge.

The Coast Guard arrived, in the person of Lieutenant Roy C. Lewis, chief of the Marine Safety Office's Department of Investigation, and his assistant, Lieutenant Peter Popko. Clean-cut, businesslike, dressed in pressed blue uniforms and projecting an air of serious authority, Lewis and Popko would check every aspect of *Summit Venture*'s machinery, the steering gear, the gyro compass, the radar, and the engine order telegraph, to determine if anything had gone mechanically wrong with the ship. With one investigator in the wheelhouse and the other stationed belowdecks in the engine room, they tested the accuracy and timing of the wheelhouse controls. They removed the onboard course recorder for laboratory evaluation.

In due course Dewey Villareal and Carl Nelson, attorneys for the ship's owners, were brought on board and escorted upstairs.

The media had caught the scent, and helicopters began to circle, wasplike, over the scene. Once they had shot footage of the appalling, shattered carcass of the Sunshine Skyway, their cameras were trained on the wings of *Summit Venture* wheelhouse, for any sign of a perpetrator.

"That," Atkins said, "was when you wanted to go out on the wing and throw up. That didn't happen, but it sure felt like it. My first reaction is 'How do we explain this?' And then you say 'You know what? There isn't any explaining. It is what it is.'"

Atkins did in fact venture onto the port wing, before the arrival of the Coast Guard investigators. He considered himself a practical man, and like all practical men he tried to see the situation from every possible angle.

"I'm standing on the wing thinking *OK, Bruce, were you at fault here? Is there anything that you did here that you need to bear responsibility for? What part did you play in it? If there was something that you could have done, should you have done it?* That's more of the seafarer's professional analysis. Because I'm an apprentice pilot, and you have the pilot, and you have the ship's master, and sometimes there may be some things that you could have done, but you shouldn't do—because you have that chain of command. And you have to respect that somebody is in charge.

"So there's the could'ves and the should'ves. If you had done one of the should'ves, would that have made any material difference? Once you come to grips with that, then that gives you the perspective on how you need to deal with the situation going forward."

In those few moments Atkins decided that Lerro had done everything possible to avoid catastrophe, and that no one—pilot, master, or apprentice—should be held accountable for something that was taken out of their hands.

Next up the gangway was C. Steven Yerrid, whose law firm Holland and Knight represented the Tampa Bay harbor pilots, accompanied by two other attorneys from the firm's admiralty department. They bounded up the stairs and burst into the crowded wheelhouse, out of breath but nevertheless all business.

Yerrid, who'd never met his new client before, introduced himself to Lerro and immediately began the process of protecting him. No, he told investigator Popko, there would be no questions today, although of course the pilot would be made available soon enough. "Have you talked with Bruce yet?" Lerro asked his attorney.

Until that moment, Yerrid had had no idea there had been a second pilot aboard *Summit Venture*.

He hurried Lerro and Atkins out of the wheelhouse, down the zigzag stairs, down the pilot ladder, and into the waiting pilot boat *Egmont*.

With the local news choppers buzzing overhead, the light 50-foot craft shot across the bay, through waters now disquietingly calm.

"An eerie quiet enveloped everyone on board," Yerrid would later write. "It was the silent passage of moments that could not, would not, tolerate anyone's interruption.

"And no one said a word. Not one."

Florida governor Bob Graham happened to be in Tampa on May 9, on government business. Aboard the Coast Guard vessel *White Sumac*, he toured the site within hours of the collision.

"At that point, human rescue was the focus," Graham recalled. "The busload of people had fallen into the bay, and no one was sure how many others might have fallen in. It wasn't about thinking about the bridge, or what might happen in the future, it was a minute-to-minute operation attempting to rescue as many people as possible. Or at least to find the bodies of those who were beyond help."

Graham's brief visit was featured in all the news reports that evening and throughout the weekend. It was somehow reassuring for

Floridians to watch footage of their governor standing in his shirt-sleeves on the deck of *White Sumac* as the buoy tender circled the disaster scene, concern etched into the contours of his boyish face.

It was one photo op Graham would gladly have done without. Between the *Blackthorn* disaster, the Mariel Boatlift crisis, and Miami's Liberty City riots, 1980 was shaping up to be a particularly dark year for Florida. And it was only May.

The Tampa pilots were all reminded of their own vulnerability and mortality. "It could have happened to any one of us," said Gary Maddox. "I think every one of us who was honest about it thought that. Given the circumstances that he was trapped in.

"We were more cognizant of the fact that we had engine failures and rudder failures, and we were afraid that that would be the downfall, more than something that was under power. But the intensity of that storm, the way it came in and the way it was masked, it was very hard to detect and anticipate.

"Before, we always considered 'Well, you can slow down, or adjust your speed, do something.' You're never going to think to yourself 'Somehow, I would have made it.' Because you don't want to deal with the reality that maybe you wouldn't have."

Under the banner headline "Skyway Span Falls," the front page of that afternoon's *Evening Independent* was devoted entirely to the incident. It was the first report in the print media.

"I felt it; I saw the superstructure falling," witness Jay Hirsch told the paper. "The first thing I did, I looked for heads bobbing. There were none."

A toll collector on the Pinellas side said that a line of about a dozen cars coming from the south had sped by her booth. Several made a momentary stop, and the drivers delivered a chilling message. "They were shouting to us that the bridge was gone, that the bridge caved in," Rae Duato reported. "We had a hard time getting them calmed down to tell us what happened."

Ironically, on the blackest day in Tampa Bay history, the *Evening Independent* did not need to make good on its pledge to distribute the paper for free; after the early storm left havoc, the sun *had* come out.

10

Red Alpha Zulu

The phone rang in Bill Covert's St. Petersburg home just as the director of Eckerd College's student dive and rescue unit was getting ready to leave for campus on an overcast Friday morning. The urgent call came from Coast Guard dispatch in Pinellas County—a "boat," the voice explained, had hit the Skyway Bridge, and the Eckerd crew was being summoned to the site. That was all the information the dispatcher had.

"To me, a boat hitting a bridge means a small boat hit a small concrete bridge," Covert recalled. "Probably severe damage. Maybe people hurt."

The team—consisting of student divers with an average age of twenty—was always on call and was required to be ready to take to the water within twenty minutes. Covert reached them all and instructed them to rendezvous at the college docks. The divers were instructed that they were dealing with an incident coded *Red Alpha Zulu*. Red meant life-threatening, Alpha was extreme danger, and Zulu indicated divers required.

Covert had begun the program in 1971 to render assistance to the school's recreational watersports teams, which practiced on the Pinellas side of Tampa Bay. The rescue team ran crash boats for Eckerd regattas, monitored safety standards, and essentially kept the other students from drowning, or worse.

Eventually public boaters with failed motors or other such problems began to hail the Eckerd vessels as they passed. Covert realized he could turn this into a new student development program, and six years after it had been started, the school's rescue team was officially recognized and accepted by the Coast Guard, which did not have its own team of divers. The Coast Guard's search and rescue efforts were limited to boats and boaters on the surface.

By 1977 the Eckerd dive and rescue team was regularly recruited to recover the bodies of suicide victims in the waters around the Skyway. On several occasions they had been dispatched to the wreckage of small planes that had crashed in the vastness of the bay. And, of course, Eckerd divers had been involved in the *Blackthorn* rescue.

Most of the work, though, was nickel-and-dime stuff: recreational boaters who got stranded or ill and needed aid. Sometimes a small vessel clipped one of the Skyway supports, but those incidents were uncommon, and there had never been serious injuries. That was what the thirty-three-year-old Covert was expecting when he rallied his first two responding divers in the school's 26-foot dive boat *Rescue 2* on the morning of May 9, 1980. *Rescue 4,* slightly smaller, followed with another diver, along with a scuba-certified Pinellas County firefighter who had heard the call and volunteered to help.

As Covert's craft rounded nearby Maximo Point, which led to the open bay and the towering Skyway, he and his divers saw it. The Skyway was a familiar sight, a landmark they dealt with every day, and there it was, right where it was supposed to be. But something clearly was amiss. The giant was . . . hobbled.

At that precise moment, the radio crackled with updated information: Be advised, said the voice. It wasn't a small boat that had hit the bridge; a freighter had "plowed into it."

"It was very surrealistic," Covert said. "You have the bridge down, girders in the water. The ship was up against the bridge with a tug trying to push it away.

"And the sun was out and it was a beautiful morning, and it was deadly silent. No noise, no nothing. Just the sound of the tug pushing on the ship. We were like, 'Oh my God.' It was just horrific. To look up and see that and *then* get the radio call 'The bridge is down, the bridge is down . . .'"

Covert and his crew motored on and went down the west side, past the maimed main span and past *Summit Venture*, hulking and still, looking like a big dog that had just soiled the floor but didn't understand why it was suddenly the center of attention.

They surveyed the scene, taking note of several helicopters circling in deliberate fashion over the water on the Tampa side of the bay. It was obvious the chopper pilots were looking for survivors or bodies that might have been swept eastward with the incoming tide.

Cutting under the bridge, mouths open in disbelief as they beheld the yawning split 150 feet up, the Eckerd crew were hailed by people aboard another small craft. It had been commandeered by a reporter and photographer from the *Evening Independent*, who were shouting, waving their arms and pointing down at the water, at the area directly under the south span, where the road and the superstructure had splashed and sunk.

"We looked where they were pointing, and we could see the wheels of the Greyhound bus bobbing in the water," Covert said. "The bus was being kept afloat by the buoyancy of the wheels. So we headed right over there thinking 'My God, what if there are people trapped in that bus?'"

Keeping clear of the mangled girders, the Eckerd boats dropped anchor in the channel. They were directly under the section of gridwork that hung by a thread, with Dick Hornbuckle's big yellow Buick balanced at its edge, literally at the end of the road.

As his divers pulled on their gear, Covert pointed upward and quietly issued a warning: Be aware of that thing, guys. We don't know how stable it is. "If it starts to fall, just jump for it," he told them. "That's the only chance you have."

Within moments the young divers had gone overboard and proceeded carefully to the upside-down passenger bus. The flood current was strong. They hoped there were air pockets inside the bus, which could mean survivors.

Mike Rosselet was twenty-one years old and Covert's most indispensible assistant. He was both the team's dive master and the designated captain of *Rescue 2*. "You always think you're going out to save a life," Rosselet explained. "And in this case, we thought the same thing. But once we got out there we realized it was just a matter of recovery at that point."

Rosselet was the first diver in the water. He quickly discovered that the bus, which had inverted as it fell and was facing north—the opposite way from its original direction of travel—was caught on the wreckage of the steel superstructure resting on the channel bottom. "You could stand on the bottom of the bus," he said, "and almost be awash."

He and another diver found a way in between what remained of the passenger door and the windshield. The body of Michael Curtin, the bus driver, was the first to be recovered.

"The bus was pretty compressed," said Rosselet. "The whole top was compressed and you only had so much access. You were just kind of reaching in and pulling out what you could pull out. There was really no bus to go into, because it was so damaged."

The divers had a good ten feet of visibility—the fast-moving weather cell had long since moved on—but staying in one place was tricky because of the powerful current flowing through the deep channel.

"You were really just looking into a tangled mess," Rosselet recalled. "You'd see bodies in little pockets and extricate what you could. But there were no air pockets. That thing was so damaged that there was no chance of survival."

Divers from the fire department rescue squad and the Florida Department of Transportation soon joined in the grisly hunt. Seven bodies were pulled from the bus. No one could guess how long the wrecked coach would stay afloat, so the crew worked quickly, swimming the bodies to the dive platforms at the stern of *Rescue 2* and *Rescue 4*. On board, the others would then hoist them onto the deck and zip them into black vinyl body bags.

Firefighter Gerard Chalmers was the first to dive to the bottom of the shipping channel. "One of the first things I saw was a pile of rivets, like an anthill," he said. "They had been sheared off. The force it would have taken was stunning."

Chalmers came upon a crushed El Camino. He could see the victim inside, but retrieval—with chunks of shattered concrete and girder pinning the wreckage to the bottom—was clearly impossible at that time. He took the dead man's briefcase as an identifier, marked the spot, and swam on.

"By this time, Coast Guard was on the scene," said Covert. "They had an 82-footer, and they became on-scene commander. You had fire department personnel up above. I could see fire trucks and rigs. The tug was still pushing on the ship. There were other press boats, other agency boats.

"I stopped at one moment and I looked around. It was like the entire world was watching us as we brought up one body after another."

Volunteers came from all over the bay. Fishing boats from the Manatee side began netting personal items—purses, shoes, suitcases—as they bobbed to the surface and began to float away with the current.

Lieutenant John Sixbey of the Hillsborough County Sheriff's Office was one of several volunteer divers. "The current was so violent, and it was traveling at such a fast speed," he said. "No one could hang on to the superstructure with one hand; we had to hang on with two hands."

Sixbey and deputy Terry Longpre reported about four feet of visibility in the water, making it necessary to move carefully along the jagged bridge wreckage. "The rebar was vibrating so violently, it looked like spaghetti in the water," said Longpre. "At the same time, you could hear the superstructure creaking."

As the divers grimly continued their task, Pinellas Fire Express unit chief Jerry Knight and his crew were considering Hornbuckle's car, still perched precariously at the edge of the fractured grating. A tow truck had been summoned from Harold's Truck Stop.

Hornbuckle himself was still on the concrete slope of the bridge, giving his statement to police, and answering questions from the gathering crowd of reporters, who were allowed access to the site for most of that first day.

Ken James, once he had finished giving his own statement, had graciously offered to drive Hornbuckle's shaken passengers back to the Apollo Auto Sales office, where their own vehicles were parked.

Knight, like Covert 150 feet beneath him, was concerned that the weight of Hornbuckle's car might be just enough to drop the heavy section of steel—and the big Buick itself—into the bay. But he was not about to let tow driver Bob DeMond, or his truck, anywhere near the teetering grate—for which DeMond, a mechanic who had been pressed into service because the regular driver was out on another call, was grateful. Knight ordered his men to attach a large hook and chain to the car's rear axle and delicately pull it back from the brink.

"They walked down there with the chain, got to the car and dropped the chain on the steel grate," recalled Covert. "Well, we're directly below, and the noise of that chain hitting that grate . . . we thought 'Oh my God, it's letting go, it's coming.' There was one just terrifying moment."

Bob DeMond gave Hornbuckle a ride back from the Skyway in his tow truck. "He was pretty nervous," the mechanic later told reporters. "I remember he wanted to stop at the store so he could get some cigarettes because he had smoked all of his. He asked if he could buy me some, but I don't smoke."

By 10:30 the two college boats were riding low in the water. The crew had laid out the seven bodies, tucked into bags, on the deck. "Unless we had stacked them like cordwood, that's all we had room for," Covert recalled. They were not equipped to handle that much raw devastation. Once the decks were filled, the last two bodies had to be strapped indelicately to the diving platforms at the stern.

The local media hungrily snapped photos from their chartered pleasure craft. There had never been anything so terrible in all of Tampa Bay's history.

A temporary morgue had been set up at the Fort DeSoto fishing pier on Mullet Key, a straight four-mile shot across the water. As the closest dock, it was a logical staging area, something the various

Fig. 17. Two days after it inverted and hit the water from 150 feet, Greyhound No. 4508 was exhumed from the shipping channel on May 11, 1980. Courtesy of *Tampa Bay Times.*

rescue units had discovered during the aftermath of the *Blackthorn* tragedy in January. That had been a grim dress rehearsal for *Summit Venture.*

The horizon was beginning to darken again; Covert could smell rain on the way.

Blood from the broken bodies was mixing with seawater as it came in through the scuppers of his anchored rescue craft, sloshing back and forth as the impatient waves rolled underneath.

A Boston Whaler with divers from the Department of Transportation had arrived, and the crew tied off to the back of Covert's 26-footer. Their job was to inspect the piers of the northbound Skyway, beneath the surface. The wheels of practicality were already turning: before DOT could consider reopening the northbound span to two-way traffic, the supports had to be inspected for damage that might have been inflicted by sections of the southbound span as they fell, or by the errant *Summit Venture* itself.

After some tense territorial squabbling between the Eckerd and DOT crews, the Coast Guard intervened. Covert's boat was freed up, and the student divers left the scene.

A crowd had gathered at the Fort DeSoto pier; behind hastily erected police barricades dozens of people strained to get a look at the incoming dive boats. To Covert's disgust, he saw gawkers hoisting small children onto their shoulders for a better view of the broken bodies. Local TV cameras were there to capture it all. Scenes of bloodied bodies being awkwardly hauled off the boats and onto the dock were beamed live into Tampa Bay homes.

Once the Eckerd team had unloaded its sad cargo, Covert ordered the rescue boats back to their home docks, to be sprayed clean of the blood wash and disinfected.

Just then Covert was notified that the search had been suspended for the rest of the day, as the unsmiling sky was turning that early-warning charcoal color again, and more wind and rain were predicted. And somebody had made a passing comment about the sharks that commonly cruised the deep and fast-moving water of the shipping channel.

At this early stage, no one was sure how many vehicles were submerged or how high the body count would go.

* * *

From his cot in Sick Bay Wes MacIntire could hear the media helicopters circling over *Summit Venture*. In his delirium he assumed one of the choppers was coming to evacuate him, along with the other survivors who had to be somewhere else on the ship. To add to his confusion, the only men with whom he had communicated spoke either rudimentary English with heavy Asian accents or no English at all.

When he came aboard to take control of the ship, Park had checked in on MacIntire and found the bruised and shaken man babbling incoherently but physically more or less intact. Park and Liu discussed how to get the sole survivor off the vessel; it was already clear to them that there would be no others.

"That ship was high," Park said. "So I told them to get two volunteer crew members, and call the Coast Guard boat over. At the waterline, lower the lifeboat down even with the deck, put the man on a stretcher and put him in the lifeboat, lower the lifeboat down with two attendants, put him on the Coast Guard cutter and take him to the hospital in St. Pete."

Wes MacIntire had been on board *Summit Venture* for over two hours. The fresh air, moist and salty, filled his lungs. But he was severely disoriented.

The journey to Mullet Key was brief, but the bouncing was rough. "I didn't break my back in the accident," MacIntire shouted at his rescuers on the cutter, "but goddamn it, you're going to break it taking me in!"

To shield their charge from the further indignity of getting rained on as he bounced along, helpless, in the rescue boat, someone had thoughtfully covered MacIntire with the next-closest thing to a waterproof blanket: a body bag, which they wrapped around and under him, tucking it gently underneath his chin. A makeshift white bandage covered the top third of his head.

When the cutter was unloaded at the Fort DeSoto fishing pier, the two men who received his wooden stretcher began to carry him toward the coroner's step-van, open at the back, with the intention of loading him in with the other unfortunates.

MacIntire was able to move his elbow and groan, at which point his rescuers cried out in astonishment—"Oops! This one's alive!"—and U-turned, double-time, for the ambulance that waited nearby, its engine running.

Although Palms of Pasadena was the nearest hospital, MacIntire insisted on being taken to St. Anthony's, downtown, where he could be seen by Edgar Buren, the doctor who had reattached his nearly severed finger following the roofing mishap of a few months before.

Buren told the media that MacIntire's injuries were not life-threatening; his neck was possibly sprained, he had some seawater in his lungs, and his right foot was black and blue (the result, as it turned out, of Wes slamming it down on the Courier's brake for the freefall into the side of the ship). Buren stitched up the gash over his eye.

"I knew God was with me on this one," Wes told the doctor, who canceled all his afternoon appointments to tend to his friend. "But this morning, God gave me a hell of a ride—and got me all wet."

A throng of reporters was allowed into the hallway at St. Anthony's as MacIntire was being rolled out of the emergency room. They surrounded the bed, and in a low whisper he recounted the story as well as he could remember it. The impromptu press conference lasted all of two minutes; as the elevator arrived to take the patient up to his private room, the word went out: there was *one* survivor!

Wes MacIntire became "The Man Who Wouldn't Die" and "The Man with Nine Lives." His survival was the only good news that day. Buren told the papers it was a "miracle."

Betty had spent the morning glued to the television. When the news of the bridge accident first broke, she had called Wes's boss at Palmetto Meat Dispatch, only to be told that her husband hadn't shown up for work. She gave in to her mounting wave of dread. She cried. She prayed.

A neighbor drove her down to the Skyway, but by the time they arrived the police were not letting anyone out onto the bridge. Betty watched helplessly from the causeway more than two miles away as the search-and-rescue work went on—wondering if her beloved husband was down there among the twisted steel and concrete. "I figured 'What better way could God take him?'" Betty said. "He loved his truck, and he loved the water."

When the wind and rain picked up again, and the search was called off for the day, Betty and her friend went back home to wait by the phone. There was nothing else to be done.

It was about 12:15 p.m., nearly five hours after the accident, when St. Anthony's called to tell Betty her husband was alive and in their capable hands. By then the MacIntires' daughter, Donna, had already boarded a plane in Massachusetts, expecting to hear the worst once she stepped off at Tampa International Airport.

Betty arrived at the hospital and rushed through the mob of reporters congregating in the hallway. Upstairs she found Wes in bed, groggy from painkillers, a fresh white cotton bandage taped over his left eye.

"I hit my head again," he said to her, smiling weakly, as she walked through the door.

Donna arrived in midafternoon. "He was awake, but we wouldn't let him watch TV because my mother didn't want him to know that anybody had died," the MacIntires' daughter said. "He was talking to us about hitting his head again. And he said 'When they pull the truck up, I'm going to hose it down real good and get all the salt out, and it'll be fine.' Of course, we already knew that a lot of people had died.

"We came back the following morning, and he was in tears. He'd put the TV on and was listening to the news. He said 'I was looking for people; I could have helped somebody.' He was just a wreck after that. That hit him real hard."

* * *

According to the National Transportation Safety Board, a vehicle falling 150 feet impacts the water at about 67 miles per hour. This is commensurate, the NTSB said, "to impacting a semisolid object at a similar speed."

It was estimated that the plunge lasted slightly less than two seconds. Which meant that the twenty-six occupants of the Greyhound bus, and the nine people in five automobiles that followed Wes MacIntire to the Skyway summit, had no time to react and probably did not know what was happening to them.

The Greyhound coach was removed from the water, by crane, on May 11. "The bus had almost bent double from the impact," said highway patrol trooper Charlie Wells. "As they pulled it up, there were bodies floating out with the bus and flowing with the tide. All the boats of other responders were picking up the bodies as they flowed out, and you could see the luggage in the bus floating away."

Because the shipping channel and the center of the Skyway were technically in Hillsborough County, that county's medical examiner received the bodies and conducted the autopsies. The official report: twenty-eight victims died from blunt trauma injuries. The other seven drowned.

They were just ordinary people, residents and commuters, each probably apprehensive about the morning's ghastly weather. But like Wes MacIntire, Don Albritton, Dick Hornbuckle, and the others who tempted fate, they were probably also familiar with the fickle nature of spring storms in Florida and decided to press ahead. Each had somewhere to go. And their instincts told them that the Skyway would deliver them, as it always had, safely to the other side.

Every ten days, seventy-three-year-old Hildred Dietch made the trip to Bradenton to get her hair done; her preferred hairdresser had relocated there from St. Petersburg. Her husband Harry, a shoe salesman, had a business appointment in Manatee County and was giving his wife a lift in his 1975 Ford LTD, a big black and yellow sedan.

James Pryor, who lived in Seminole in northwestern Pinellas County, was production manager for Kee Manufacturing, a Bradenton company that made lawn mowers. It was the forty-one-year-old Pryor's briefcase that firefighter Chalmers had recovered from the shattered remains of his white and tan 1976 Chevrolet El Camino, pinned under bridge debris.

John and Doris Carlson were looking forward to a weekend sales convention in Miami in their silver 1980 Chevrolet Citation. He was a traveling tool salesman. She was a nurse in the cardiac care unit at Clearwater Community Hospital. The car had somersaulted during its freefall from the summit and was inverted when it struck the water. The Greyhound bus—the last vehicle to leave the broken bridge—landed on top of the Carlsons' car. John Carlson's son Jeff, from a previous marriage, would graduate later in May from the U.S. Naval Academy at Annapolis.

The other vehicles were a light green 1979 Chevrolet Nova, driven by Les Coleman of St. Petersburg; a light blue 1980 Fort Granada, driven by Charles Collins of Tampa, and a silver 1979 Volkswagen Scirocco with Robert and Delores Eve Smith of New Jersey inside.

The Scirocco and the El Camino, it is believed, were on the bridge as it fell. They were several car lengths behind Wes MacIntire's pickup. The other vehicles fell from the break, over an estimated two-minute period following *Summit Venture*'s impact with pier 2S.

Robert Smith's body was the last to be recovered. On May 14, five days after the accident, it was found floating nearly three miles east of the Skyway, on the Tampa side.

Thirty-five people, gone with the wind:

Duane Adderly; Miami, Florida
Alfonso Blidge; Miami, Florida
Myrtle Brown; St. John's, Newfoundland
Willis Brown; St. John's, Newfoundland
John Calloway; Miami, Florida
Doris Carlson; St. Petersburg, Florida
John Carlson; St. Petersburg, Florida
Leslie Coleman; St. Petersburg, Florida
Charles Collins; Tampa, Florida
Michael Curtin; Apollo Beach, Florida
Laverne Daniels; Miami, Florida
Sandra Louise Davis; Gainesville, Florida
Harry Dietch; St. Petersburg, Florida
Hildred Dietch; St. Petersburg, Florida
Sharon Dixon; Miami, Florida
Brenda Joyce Green; Miami, Florida
Robert Harding; Glenn Falls, New York
Gerda Hedquist; Port Charlotte, Florida
Aubrey Hudson; St. John's, Newfoundland
Phyllis Hudson; St. John's, Newfoundland
Louise Johnson; Cataula, Georgia
Yvonne Johnson; Miami, Florida
Horace V. Lemmons; Kings Mountain, North Carolina
Lillian Loucks; Winnipeg, Manitoba
Louis Lucas Jr.; Mobile, Alabama
Margurite Mathison; St. Petersburg, Florida
Tawanna McClendon; Palmetto, Florida
MaNesha McGarrah; Tallahassee, Florida
Wanda McGarrah; Tallahassee, Florida
Ann Pondy; Winnipeg, Manitoba
James Pryor; Largo, Florida
Mel Russell; Chicago, Illinois
Delores Smith; Pennsville, New Jersey
Robert Smith; Pennsville, New Jersey
Woodrow Triplett; Bainbridge, Georgia

11

The Court of Public Opinion

"How does it feel to have a murderer for a father?"

The woman's voice on the telephone sent a chill through thirteen-year-old Charles Lerro. He'd just come home on the bus from Buchanan Junior High School, where the classroom intercom had been buzzing all day with Skyway updates, gleaned from the frantic live coverage by all the local radio and TV stations.

Eventually the boy had discovered that his father had been the pilot aboard *Summit Venture*, and when he arrived at the house on Wayne Road, there was no one there to greet him, no one to explain what had happened. Charles was used to coming home to an empty house, as both his parents' jobs required them to be away for long stretches. He was a latchkey kid. Rarely were all three of them home at the same time.

The phone was ringing when he walked in the front door. "The first call," he recalled, "was a reporter. And the next. They were asking a lot of questions, which I did my best not to answer."

Then came the chilling query from the woman, who did not identify herself. "After that, I stopped answering the phone," Lerro said. "Her comment didn't bother me, because my father had taught me by example not to worry about what other people said or thought."

C. Steven Yerrid, of course, was paid to be sensitive to what people said or thought about his client, and he knew, without thinking too hard about it, that he had to get Lerro and his family out of the line of fire. He knew the media, and he knew what was coming.

As soon as Charles and Sophie could be rounded up—before another reporter could corner Lerro's impressionable young son or badger his wife with questions and accusations—Yerrid had the family checked into Tampa's Airport Motel under assumed names. The attorney instructed them not to answer the door, not to answer the telephone, and not to trust anyone they might encounter.

Within hours of the collision the name John Lerro was all over local talk radio and TV news. His history as a ballet dancer became sniggering fodder. Around the office water cooler and at the neighborhood tavern he was tried, convicted, and executed before the first sunset. How in the hell could anybody hit a stationary object as big as the Skyway Bridge?

"I never dreamed they would make him out to be a drunk," Yerrid said. "He rarely if ever drank. Or make him out to be a 'queer.' That was the name at the time, a slanderous name, like 'look what a piece of trash he is, he's a drunk and a queer.' It was a different world." The Lerros stayed at the motel for three weeks and ran up a huge room service bill because they rarely left the premises.

On Saturday, the day after the accident, the *NBC Nightly News* began with a cold opening. A somber Jane Pauley had shocking news for the country:

> From Tampa Bay, Florida, the revelation today that the pilot of the freighter that crashed into a bridge, killing thirty-two people, had bumped into the same bridge earlier this year and was in fact under investigation for the incident.
>
> Divers continued their grim search for bodies. A bus and at least two vehicles plunged into the bay yesterday when the freighter struck the Sunshine Skyway Bridge.

Under media pressure, the ten-member Board of Pilot Commissioners had released the information that Lerro had been investigated for seven previous incidents since his arrival in Tampa in 1976, including *Jonna Dan*, and six others that did not involve the Skyway. All three television networks, and the Associated Press and United Press International wire services, duly reported this "revelation."

Most failed, however, to mention the additional comments by the board's executive director Jane Raker: that Tampa, as the busiest port in Florida, received many more incident reports than all of the other ports combined; that the pilots themselves voluntarily make the reports; and most important, the board's conclusions that five of the seven Lerro incidents were unavoidable and not Lerro's fault in the slightest—they had resulted from steering-system failures, unfendered docks, or bad weather-related groundings.

The other two were more extensively investigated. In the *Jonna Dan* case, the Marine Casualty Review Panel commended Lerro for "avoiding a more serious collision"; the December 1979 *Straits of Canso* accident, in which Lerro's ship broke a gantry crane while pulling away from its mooring in port, was ruled the result of both weather issues and the erratic movement of tugboats.

Even more significant, during the three years that Lerro had been employed in Tampa, pilot Harry Williams had eleven reported incidents. Thomas Baggett had ten and was fined, in two cases, by the Coast Guard. Fred Enno reported seventeen incidents, and the Coast Guard suspended his license for two months.

Having reviewed 236 accident reports statewide in its five years of existence, the Board of Pilot Commissioners had never removed or disciplined a pilot. It was a dangerous profession, and an inexact science, and no two sets of circumstances were ever the same.

In the Williams and Baggett cases, board member Ronald E. Schaefer alone voted to discipline the pilots; he was a Pensacola harbor pilot and an ex-Navy man.

Schaefer had issued a warning to the other board members the previous November: "There appears to be a high incidence of casualties among a small number of pilots," he wrote. "I suggest that we investigate some of these casualties and revoke the license of any incompetent pilot before a major catastrophe results." In the wake of *Summit Venture,* the board came under intense criticism for what was perceived to be reluctance—or out-and-out inability, for whatever reason—to rein in its own. In effect, John Lerro had all but volunteered for scapegoat duty.

Later that first Saturday evening after the Skyway collapse, much of America tuned into NBC's *Saturday Night Live,* which—despite the recent departure of founding stars John Belushi and Dan Aykroyd for greener show business pastures—was still one of the highest-rated shows on TV.

Three minutes into the "Weekend Update" segment, a photo appeared onscreen of Dick Hornbuckle's Buick, perched precariously at the jagged edge of the Skyway, and newscaster Jane Curtin read:

> The St. Petersburg highway department had to collapse the Tampa Bay Bridge yesterday, in order to apprehend a car that drove onto the bridge without paying the toll. The department apologizes to other cars and buses that were inconvenienced by the incident.

"We of the Tampa Bay area have been in a profound state of shock over the collapse of the Sunshine Skyway Bridge Friday," Clearwater resident Martha Johnson said in a letter to the *St. Petersburg Times.* "Saturday night I listened to the 11 p.m. news and was not in the room when it ended, so *Saturday Night Live* was on for a few minutes. I could not believe my ears when they made a joke about the collapse of the bridge. I find it incredible that anyone could find it funny. Perhaps they would feel differently had a member of their family been involved."

"Can death ever be funny?" asked *Times* TV columnist Robert Bowden. "How about abortion? Rape? Incest? The Jewish holocaust? No. These realities can never be transformed into jokes acceptable to some persons." Bowden went on to defend "sick" and "avant-garde" humor, saying that it, too, had a time and a place. "But yes, Mrs. Johnson, I too think the Skyway joke was in poor taste, for this reason: There is a time to laugh and a time to cry. This time, NBC confused the two."

The next morning the original 1954 span was reopened to two-way traffic, following a hasty Saturday-afternoon paint job on the center dividing lines. Driving over the pinnacle became a surreal experience, with the grotesquely maimed twin span at eye level just 100 feet away. The massive section of latticed steel gridwork that had supported Dick Hornbuckle's car would dangle there for a week before salvage crews blasted it free. Rubbernecking was hard to resist, even at 150 feet up, and traffic going both ways on the functional bridge slowed to a morbid crawl.

On Monday a temporary shipping channel was green-lit between piers 1N and 2N, allowing vessels limited to a maximum draft of 23 feet to pass under the broken bridge and proceed through its undamaged twin. Because the gap was just 360 feet—considerably narrower than the main passage, now blocked by debris from the fallen Skyway—vessels were allowed through only one at a time, with no passing, and only during daylight hours. This was the same temporary channel that ships had used following the *Blackthorn* incident. Lerro, piloting *Jonna Dan,* had clipped the Skyway while attempting to navigate this slim passageway.

Tampa's Port Authority breathed a loud and very public sigh of relief as vessel traffic, both into and out of the port, had been at a standstill since Friday morning. More than $1 million per day was lost as nothing moved.

Governor Graham declared a state of emergency for Pinellas, Hillsborough, and Manatee counties, in effect asking the federal government to help defray the enormous expense of the salvage operations and eventual rebuilding of a fully functional Sunshine Skyway. William Rose, Florida's secretary of transportation, planned to fly to Washington and meet with the head of the Federal Highway Administration to begin sorting out exactly what sort of bridge to build.

Florida House Speaker Hyatt Brown announced the formation of a special legislative committee to investigate the Skyway accident. Committee member Peter Dunbar of Dunedin told reporters he was particularly concerned with Tampa's harbor pilots. From what he had read, Dunbar said, Lerro "should have been disciplined and pulled off a long time before this." Dunbar expressed concern that the pilots were "being regulated substantially by the profession itself" and said he wanted to understand "why the law didn't work" to prevent the disaster. "This isn't the only pilot or the only incident that needs investigating," he said.

"Being a sailor who has been under that bridge perhaps 200 or 300 times," declared Rep. Dennis Jones of St. Petersburg, "it is absolutely inconceivable to me that pilots with today's navigational ability, radar and the basic knowledge of traveling that channel, hundreds of times, have not only had so many mishaps, but so many near mishaps we've never heard much about."

It fell to Hillsborough state attorney E. J. Salcines to consider bringing Lerro up on criminal charges. That was certainly what the public wanted, and what the anguished pilot expected.

"E.J.'s got a population of people who are demanding and crying out that somebody be lynched," Yerrid recalled. "Especially this drunk, gay guy! The property of the state—$200 to $300 million dollars—having been taken, thirty-five lives having been lost, somebody by God has gotta pay! Somebody's gotta be blamed. It's high time we had this worthless judicial system do what it's supposed to do and indict this bastard for manslaughter!"

But Salcines, instead of bowing to extreme political pressure, concluded that whatever had gone wrong, John Lerro would not have to answer for the accident in criminal court.

"E.J. called me at my house and said he'd looked into it, he'd been at the morgue, he'd seen the bodies," said Yerrid. "He understood everybody's anger. He understood everybody's thirst for lynch party justice, but that wasn't gonna happen. Because he'd seen nothing that would allow him to bring anything before a grand jury."

Yerrid told his client they had won a significant battle. The war, however, had yet to be fought.

The Coast Guard and the National Transportation Safety Board, in tandem, would conduct a series of public hearings called a Marine Board of Investigation. The objective was to gather testimony from all pertinent parties, review all available technical data, and publish a conclusion as to the probable cause of the accident. Its purpose was

Fig. 18. John Lerro at the Marine Board of Investigation hearings, Tampa, May 15, 1980. This was the first time the public got a look at the man who had destroyed the Skyway. Advising him were C. Steven Yerrid (*left*), whose law firm Holland Knight represented the Tampa harbor pilots, and the firm's admiralty specialist Paul Hardy. Courtesy of *Tampa Bay Times*.

fact-finding only; any disciplinary action or litigation would have to come from another entity.

Washington, D.C.–based Captain Edward R. Grace was one of three Coast Guard "traveling inspectors" who were assigned to trouble spots around the country. He volunteered for the Tampa case and was named presiding officer in the Matter of the MV *Summit Venture* and the Sunshine Skyway Bridge Collision.

Two NTSB lawyers were present, alongside attorneys representing the ship's master and operators, the Florida Department of Transportation, the Board of Pilot Commissioners, the Department of Professional Regulation, and Greyhound Corporation. C. Steven Yerrid represented Captain John E. Lerro.

* * *

The hearings got underway at 9:00 a.m. on Tuesday, May 13, in Room 415 of the old post office building on Florida Avenue in Tampa, which was being used as a temporary federal courthouse (renovations on the regular courthouse would not be completed until 1984).

Captain Liu was the first witness, explaining under initial questioning that he'd joined the merchant marine in 1962, after a stint in the Chinese Navy, and held master's licenses from both Liberia and China. The fifty-two-year-old Taiwanese national had come on as master of *Summit Venture* in February, and since then the vessel had been to Japan, passed through the Panama Canal, and discharged cargo in New Orleans and Houston.

At that point *Summit Venture* had been sublet to another firm, Mitsubishi International, and directed to Tampa to pick up the phosphate bound for Korea. In broken but polite English, Liu described the events of May 9, from the time he had ordered the ballast tanks blown at the Sea Buoy and raised the anchor and proceeded to pick up Lerro and Atkins.

The weather, the captain explained, hadn't been particularly bad until *Summit Venture* had passed the lighthouse on Egmont Key and then cleared the southern tip of Mullet Key. During the transit he made frequent visits to his cabin, on the floor immediately below the wheelhouse, and spent a few minutes on each of the wheelhouse wings, port and starboard, watching the whitecaps as the wind picked up.

When visibility became exceedingly poor, Liu told NTSB lawyer Paul Ebensen, he was concerned. When Lerro ordered the anchor standby, he began to think something might be wrong. Still, Liu said, "Piloting a ship through a channel in a harbor area, it's one man's job. If I interfering with his ways, I think he might upset—he make a wrong decision. Then there will be more trouble."

Liu's testimony took the entire day; most of the questions were technical, concerning the ship's radar, running speeds and data logging procedures, and the execution of chain of command between him and the members of his crew.

Morton Good, representing FDOT, ended his redirect by getting to what many felt was the heart of the matter:

Q: Captain, when the pilot boarded the ship, could you describe his condition to us? Was there anything about his condition that caught your attention?

A: No, sir.

Q: It appeared to be normal in all respects?

A: Yes, I think so. But, anyhow, that's the first time I saw him.

Q: Did he appear to be sleepy or—

A: No, sir.

Q:—wide awake?

A: No, sir.

Q: Nothing of that nature?

A: Anyhow, he is not sleepy. I think he is quite fresh.

Q: There was no evidence to you, at least, of having, for lack of a better word, a hangover or anything of that nature?

A: No, sir. No, sir.

After this somewhat embarrassing exchange, the official transcripts show that the word "hangover" was never again uttered.

* * *

Through an interpreter, fifty-year-old Chief Officer Chan Chim Yee described Lerro ordering lookouts to the bow as the weather began to deteriorate. Using the wheelhouse telephone, Chan instructed the bosun and the carpenter, who were at that moment in the dining room, to put on their rain gear and proceed forward. Since Lerro didn't give the officer a specific reason for the order, neither did Chan tell the lookouts what they were to look out for.

Chan reported sighting Buoy 16 at 7:23 just off the starboard side, which meant *Summit Venture* was in the channel, where it was supposed to be. Lerro himself opened the starboard wing door—the doors on both sides of the wheelhouse were being kept closed because of the erratic rainfall—to see it for himself.

It was at this point that the sky exploded, as if a fuse had been lit and burned down to contact. Chan himself was ordered forward to stand by the anchor. This meant the master, Liu, would stand watch in his absence. It took the officer between three and five minutes to make his way down the narrow stairs to his cabin, pull on his raincoat, and begin the long, wet walk toward the bow of the ship.

The hearing was adjourned early so that the three members of the board could take a boat out to Egmont to examine *Summit Venture*. They wanted to walk her decks, peer through her wheelhouse glass, and attempt to feel what these beleaguered mariners had felt on May 9, when everything spun out of control.

Outside the post office building, reporters were corralled by David Rabren, a Tampa Bay pilot who was not involved in the investigation. Rabren loudly insisted that the media hear about how the Florida legislature had taken the hiring of pilots away from local groups by creating the Board of Pilot Commissioners. Lerro, he proclaimed, was "unqualified" to guide ships on Tampa Bay. Not only that, he added, several other pilots hired by the state were "just as lousy as he is."

Officially the Tampa Bay Pilots Association had been keeping mum since the tragedy. But group co-manager Cyrus Epler broke this silence to respond, on the record, to Rabren's accusations. A "vast majority" of the pilots, he told the press, were "solidly behind Captain Lerro." Rabren, emphasized Epler, was "in no way to be taken as a spokesman for the organization."

Several days later Epler's fellow association manager Gary Maddox would grant a remarkably candid interview to the *Lakeland Ledger*. The attack on his fellow pilot, Maddox said, was "without foundation. John is also experienced, and I feel the attack on his qualifications is one of personal vendetta."

Maddox was defensive of his profession, which he believed the general public had a hard time understanding. "On the surface," he said, "I guess piloting looks easy. People tend to think it's on a par with driving a semi down the highway. But roads don't silt up, and an 800-foot phosphate carrier is a far cry from the longest semi. You don't slam the brakes on a ship, either, when you're in a jam. There are no speed limits or traffic signals. Each ship, depending on its shape and engine and rudder and many other factors, has different handling characteristics. The pilot has to consider all these things, as well as currents, eddies, wind, darkness, rain and fog. Ships have to move when they're ready, not at a pilot's convenience."

* * *

Chan finished his testimony on the morning of May 15, explaining that he heard the bow of *Summit Venture* strike the pier when he was stepping carefully along the slippery starboard deck, between No. 2 and No. 3 hatches. He froze when he saw the steel trestles start to twist and plummet, then turned on his rubber heels and retraced his steps in the direction of the stern.

Bosun Sit Hau Po, fifty-nine, took the stand and testified that from his position on the bow, he saw one of the Cut A turn buoys—he did not know which—dead ahead and closing fast as *Summit Venture* continued forward.

The buoy, Sit explained through the interpreter, went past about two or three feet from the vessel, on the port side. He duly reported this to the wheelhouse over the shipboard telephone system.

With the wind and rain stinging his eyes, Sit could not discern the color or number of the buoy, nor whether it was lighted. Thus whether it was 1A or 2A—turn indicators on the port and starboard sides of the channel, respectively—was never established. Lerro believed he'd heard the phrase "buoy starboard bow," meaning he was still close to the channel, if not inside it, and chose his course change based on this information. This became a famous gray area and, ultimately, in light of the understandable confusion of those tense moments, a moot point.

The order came through to stand by anchor. The carpenter, who was unaware that the chief officer was on his way to assist, made the anchor ready. "One or two minutes" later, Sit said, came the order to drop. Both orders, he recalled, were shouted in English.

Just before the anchor was let go, Sit said, he saw the Sunshine Skyway—he hadn't been informed that there was a highway bridge in the ship's path—and it was clear to him that *Summit Venture* was going to slide between the tallest of the concrete support piers and the next shorter pier. It wasn't going anywhere near the 800-foot gap that straddled the shipping channel.

He didn't know how many shots of anchor chain ran out, he testified, because he was focusing on the imminent collision and thinking about where he could run. “When I saw the ship was going to hit the bridge, I dashed,” Sit explained. “I ducked to hide myself right next to the bits.” He pressed himself to the foc’sle deck.

When the noise stopped, Sit discovered that the anchor windlasses had broken the roadway’s fall just enough to save him from being crushed. He was, he said, “very dizzy. I was scared. My muscles jump up and down.”

The next morning, sixty-four-year-old Lok Lin Ming, the ship’s carpenter, told the board through the interpreter that he’d been going to sea for forty years and had been serving on *Summit Venture* since the first of January.

Yes, he was one of the two lookouts stationed on the foc’sle head, as ordered. Yes, he had let go the port anchor when so ordered. He did not see any buoys. He did not witness the moment of collision.

As the roadway fell, the carpenter said, he leapt over the bulkhead rail, about five feet behind the anchor windlasses, and landed on the deck. He didn’t see the bosun and did not know what had become of him.

A moment later, he said, the second mate came running down the deck and issued an order “to look at both sides and see if any unusual things like bodies or something like that.” Lok explained how he’d peered over the port side and spotted Wesley MacIntire 40 feet below and clinging like a drowned rat to a section of silver steel. He recounted the details of MacIntire’s rescue.

After highly technical testimony from Coast Guard investigator Roy Lewis, Earl Evans was sworn in. The veteran Tampa pilot had been on the laden *Good Sailor*, bound westward to the Gulf, when it passed the incoming *Summit Venture* around 7:15 a.m. at Buoy 11.

Approximately fifteen minutes later, Evans explained, “There was a terrific explosion in the ship. I was hanging over the radar. The captain of the ship was hanging on this side. It startled him . . . he reached and grabbed me to keep from falling and hollered what—what explosion. ‘Did the radar blow up or what?’”

Evans deduced that *Good Sailor* had been struck by lightning. They were at Buoy 10, just west of the Egmont Key lighthouse. It was 7:25.

"At this time (the) water was white it was blowing so hard. I couldn't estimate it. And raining horizontally practically." The wind was hurricane-force—"Sixty, seventy miles an hour at least"—for a good five minutes, Evans estimated. His radar picture was obliterated. Because it was heavy, and was steaming into the wind, *Good Sailor*'s transit wasn't adversely affected by the ambush weather. Evans held his course, and by 7:30 the worst of it had passed over.

The angry black cell was on an easterly path and coming up on *Summit Venture*.

YERRID: Now, you've been a pilot in these harbors since 1957, right?

EVANS: Yes.

YERRID: Would you have had any difficulty with these types of severe weather conditions if you had been proceeding east towards the Sunshine Skyway, instead of west outbound towards the sea?

EVANS: A different situation entirely. You'd have had plenty of problems.

John "Jack" Schiffmacher, who'd been piloting the outbound *Pure Oil,* explained that the weather had hit his ship as he was a little less than two miles east of the Sunshine Skyway—"the wind was almost instantaneous from a light breeze to fifty to sixty miles an hour"—and that he had put her on a slow speed.

With his radar useless, and visibility nil, Schiffmacher ordered a 100-degree starboard turn into the shallows. Like *Summit Venture,* his ship was light—just 10 feet of draft in front, and approximately 21 at the stern—and Schiffmacher knew he'd have ample room for safe anchorage. Best to wait the thing out and stay well clear of the inbound *Summit Venture* until he could actually see something. "It was one big blur," he told the investigating board.

The key factor here was the wind. Because *Pure Oil* was coming from the east, and was still a couple of miles away from the Sunshine Skyway, Schiffmacher could expose his ship broadside to the wind blowing from the southwest with no risk of being pushed into the traffic bridge.

As for *Summit Venture*, on the western side of the Skyway, Schiffmacher agreed with Evans's earlier assessment. With the wind pummeling you eastward, he said, "You'd be like a balloon. You'd just float with the wind."

During his turn on the stand Schiffmacher was asked, repeatedly, about his radio conversations with Lerro just prior to the collision.

When *Summit Venture* was near the Egmont lighthouse, the twenty-three-year veteran pilot confirmed, he and Lerro had briefly communicated, to confirm their eventual meeting somewhere in the vicinity of the Skyway. They hadn't yet made their passing agreement. "And then later on I broadcast to the *Summit Venture* that I was pulling out of the channel to anchor in heavy rain," Schiffmacher testified.

Under direct questioning by a different attorney, Schiffmacher added more details: "And I believe I told *Summit Venture* we were coming around Three Buoy, anchoring on the flats."

Lerro never heard this second broadcast of intent, and he therefore determined that he couldn't bring *Summit Venture* hard aport before reaching the Sunshine Skyway, because of the likelihood that Schiffmacher's *Pure Oil* would be in the channel. Lerro dared not risk a collision with an oil tanker; the explosion could destroy both vessels. It might even take the Skyway down, too.

With the weather cell directly above their heads, the pilots could hear little more than noise and static on the walkie-talkies they used to communicate with one another.

Privately, however, pilot association manager B. F. Wiltshire held a different opinion about the interaction between Schiffmacher, the veteran pilot, and Lerro, the newcomer not yet comfortable in the organization. It was Wiltshire's contention that Schiffmacher—whom he sometimes referred to as "Shifty"—never actually made the radio call. To cover his tracks, Wiltshire believed, Schiffmacher lied on the stand.

On the afternoon of May 9, the association manager—back from Texas—had called Schiffmacher at home as part of his first round of questioning. "I will never forget, as long as I live, seeing my father throw the phone across the living room," Wiltshire's daughter Judy Nunez recalled. "When he got off the phone he was so angry, because even though Schiffmacher never said 'Yes, I turned the radio off,' he never *denied* turning the radio off.

"They had known one another for so long. He never accused Schiffmacher of doing it with malicious intent—'he's too dumb,' my father said. However, he always held him [Schiffmacher] equally culpable. My father knew that man well enough to know he was dumb enough to have turned his radio off. It could have been that he turned his radio off just to save the battery."

Although Lerro and Atkins both suspected Schiffmacher's lie, they never told anyone, including attorney Yerrid. The damage was done; what would be the point? And you just didn't rat out your fellow pilots.

"My father always felt that an injustice was done," Nunez said. "You have to temper that with the fact that he didn't really have a lot of faith in Lerro's navigating ability in the first place. So therefore he could always say that it may have happened anyway, regardless."

Nunez said she was in the room when Wiltshire called Lerro, who was already in hiding with his family at the motel. "I heard my father's side of the conversation," she recalled. "He told Lerro that he knew that Schiffmacher had turned off his radio.

"He said 'In spite of the fact that I was against your hire, and the fact that I'm not really convinced that you're a great pilot, I'm very sorry for what you're going through. And the fact that Schiffmacher turned off his radio is inexcusable. And I'll make sure this information is known.'"

Wiltshire was never called to testify at any of the legal proceedings; if he did disclose his suspicions, it wasn't in open court, and the matter never came up.

* * *

While the day's drama was playing out in the courthouse, there was cleanup activity on Tampa Bay: in the morning, the "old" Skyway was briefly closed while the 36-foot grid section, a grotesque and terrible

Fig. 19. One week after the collision the shipping channel was open for business, with all Skyway traffic diverted to the undamaged span. Taken from the north (Pinellas) side, this photo does not reveal the 1,200 feet of missing bridge on the south side. Courtesy of *Tampa Bay Times.*

reminder of May 9, was blasted free. It hit the green water and was promptly recovered by barge and crane, and traffic was permitted to resume.

On the same day, *Summit Venture* was tugged back under the bridge and across the bay to Tampa's ship repair yards for patching up of the the minor damage inflicted by the impact with pier 2S and by the falling roadway, which had already been plucked off the bow by crane and taken away on a barge.

Junk dealer Max Zalkin bought the steel and concrete remains of the Skyway from the salvage company that had dredged it from the bottom of the bay. He stored hundreds of tons of the stuff, bent and barnacle-encrusted, on the fifteen muddy acres of scrap and trash he owned in Gibsontown, alongside the bay at the south end of Hillsborough County.

In June, just a month after the disaster, Zalkin announced his intention to sell chunks of the broken bridge as souvenirs, with the smallest pieces going for $1—for the larger sections the price was negotiable, but the customer had to haul it away. Bolts and rivets, shaved off and sunk when the truss and trestle had twisted and pulled apart, were encased in plastic and made into keychains, which were handed out to dealers doing regular business with Zalkin Scrap.

"Some people may be offended that we're doing this," said Zalkin, who kept a basketball-sized hunk of asphalt for himself. "But other people would be offended if they didn't have a chance to buy some."

He had discovered, inside the knots of wrecked steel, a red, white, and blue section of Greyhound 4508's back door. He did not intend to keep it. "The bridge doesn't get to me," Zalkin said. "But that piece of the bus gives me the creeps."

* * *

On Friday, May 16, one week after the collision, after the lunch break, Lerro finally made his appearance. Since harbor pilots were an exceedingly private breed, and no photographs had been circulated by the association, the media had been speculating all week on what their mystery badman looked like. Would he have horns and a pitchfork? Would he be babbling, drunken, defiant, apologetic?

Lerro took the stand around 2:00 p.m., dressed in a navy blue double-breasted suit and a striped tie Yerrid had sent over to the motel. He looked impossibly handsome with his deep, serious eyes and wavy black hair. More than one reporter noted that he bore a strong resemblance to the actor Al Pacino.

Lerro's testimony on the first day was limited to an hour before Grace adjourned. The next day was a Saturday, but the hearing was to proceed, with only the chief witness on the stand for the duration. On Friday evening his dark, chiseled good looks were splashed all over the local and network news broadcasts. One week after the collapse of the Sunshine Skyway Bridge, the world got to put a face to the already well-worn name of the perpetrator.

On Saturday morning Coast Guard investigator Lieutenant Commander David Cole began by asking Lerro to recount, as best he could, everything that had transpired from the moment he and Atkins boarded *Summit Venture* at approximately 6:30 a.m. on May 9. John

Lerro was on the stand for three and a half hours, and before the noon recess, he'd been grilled by every lawyer in the room.

"There was a lot of talk beforehand about 'are you going to stick it to him, take his license away?' Grace recalled. "But I think when he got done testifying, some of that got softened.

"He got to sit there and tell his story without anybody interrupting him. And as it went on, you could just see everybody in the place started relaxing. Even I did that; my note taking was really screwy. I had to read it over when I got done. He had an idea what he wanted to do, and he thought he had everything under control. But he didn't. And it made you think 'Maybe he really is a guy who knows how to pilot a ship, but he isn't a super-pilot.'"

Two days later, on Monday, Atkins recounted the events of May 9 from his perspective. Having completed—more or less—the thirty required bay transits, he was now officially a deputy pilot.

On the stand Atkins was professional and unflappable. He was asked, over and over and from every possible perspective, about the ship's handling that morning. About the captain and crew. About the weather. About Lerro.

On the bosun's report of a buoy sighting at the crucial turn: "The report was definitely and without reservation 'Buoy starboard bow. Buoy starboard bow'—repeated twice. It was very clear and it was in very good English."

Atkins said he'd read the bosun's account, in the newspapers, of the buoy passing along the port side of *Summit Venture*. "With the wind the way it was at that time, it's entirely possible," he said. "Obviously from the results of that morning, there was a lot going on that was totally beyond our control. It could have been.

"But if you were to ask me now, based on looking back, I would say no. I would have to assume that it still went down the starboard side."

In the seconds between impact with the Skyway and Lerro's first frantic mayday call, Atkins said, he heard a garbled radio transmission. "It did sound like Captain Schiffmacher's voice saying, well, I've decided I am going to anchor. But it was not a transmission that would have given us any indication prior to the accident that he was going to anchor.

"It was almost an after-the-fact transmission. I did hear it. I don't know over which radio it came and I don't know if it was ever acknowledged by anybody."

Atkins asked presiding officer Grace for permission to make a statement. In his ten years as a licensed mariner, he told the board, "I have never been engulfed by a storm or a weather system that had such accelerated intensity that this storm had.

"It is just my own personal opinion from being there that the other pilots that I had seen in other areas and the other mariners that I have worked with and that I have seen, that given the same set of circumstances and the same weather conditions as they existed out there at that time that morning, that I don't know of a one that would have performed in a more professional manner than I saw Captain Lerro perform.

"The elements had control of that vessel at that time, Captain. There was just nothing that Captain Lerro or anybody could have done at that time."

On Friday, May 30, *Summit Venture*—laden with thirty-one thousand tons of crushed phosphate—left Tampa for the South Pacific. Gary Maddox was piloting the ship, which was flanked by a pair of tugboats for the midafternoon pass underneath the Sunshine Skyway. Captain Liu was glad to get back on the water.

12

Guilty, Guilty, Guilty

At 7:30 a.m. scattered showers and a very few heavy thundershowers were embedded in a large area of rain that stretched across the state from St. Augustine and Titusville to Englewood and Cross City then west and southwest into the Gulf for 100 miles. Movement was to the east at 35 miles an hour.

National Weather Service radar summary, in its entirety, for Tampa Bay area, Ruskin, Florida, 6:30 a.m. EDT, Friday, May 9, 1980

A poll conducted by St. Petersburg–based Premack Research in June concluded that four out of ten Tampa Bay residents refused to drive over the remaining Skyway span.

Of the 793 people queried, 25 percent believed John Lerro was responsible for the crash. Others blamed structural faults in the bridge, poor inspections by DOT, the weather, and Captain Liu of MV *Summit Venture.*

On June 17, the day before Lerro was scheduled to return to work on the bay, the Department of Professional Regulation suspended his piloting license in a three-page emergency order. His first job on the roster, ironically, was to bring a phosphate freighter inbound.

In a public statement, DPR secretary Nancy Wittenberg said Lerro "lacks the necessary skill, judgment and presence of mind to pilot a vessel in a trustworthy manner."

After a "probable cause" hearing in which Lerro once again relayed his version of the events of May 9, three members of the state's Board of Pilot Commissioners had voted that there was sufficient evidence of Lerro being at fault, and that disciplinary action had to be taken.

Wittenberg went further, indicating that Lerro "did not properly react" to his extraordinary circumstances on May 9. "This demonstrates that John Lerro does not possess the ability to execute command of a vessel in a stressful situation," she announced, "and that should he continue to pilot vessels on Tampa Bay, lives and property will be unnecessarily jeopardized."

Yerrid immediately appealed the decision, and lost, and a full hearing in front of a specially appointed officer of the court was scheduled for October.

The young attorney believed the state was going after Lerro so zealously in order to distract the media and the public from the structural integrity problems brought to light by Art Goodale, and the fact that the Skyway had been operational for almost thirty years without any protective fendering—in other words, the DOT and the Department of Professional Regulation had been gambling with people's lives every day that vehicles crossed the bridge.

Although he knew he could never prove this as fact, Yerrid intended to suggest it in his closing arguments.

Until October, there was nothing to be done. The family returned to the house in Odessa. Sophie went back to sea for Texaco, and John stayed home to tinker with his radios, row in the pond, and look after Charles, whose summer vacation from school had just begun.

There had been a seismic shift in media attention. Over the course of the first round of hearings, the story had traveled from the front page of the Tampa Bay newspapers to the B section, the regional news. And once Lerro had shown his face in court, the national news organizations stopped covering the hearings altogether and went looking for the next big disaster story.

It didn't take long. On May 18 Mount St. Helens erupted, killing 57 people and destroying 250 homes, 185 miles of highway, and 47 bridges in Washington State. The deadliest volcanic eruption in U.S. history wiped the deadliest ship/bridge accident in U.S. history off the nation's front pages.

John Hayes, Lerro's roommate from the Maritime Academy, remained a good friend and confidant throughout.

After leaving the merchant marine Hayes earned an MBA from Columbia University. At the time of the Skyway incident he was a collision researcher at the school's Computer-Aided Operations Research Center, where he conducted extensive testing on electronic mapping systems as aids to radar. Using a ship simulator, Hayes designed and managed experiments that he hoped would help reduce the incidence of ship collisions and groundings.

"John looked to me for answers," Hayes said. "After struggling with it for a long time, to satisfy his curiosity I came up with a scenario where he could have increased the speed up to 15 knots, as he made the port turn, into the mud. That's what I would have had him try if it was not for the *Pure Oil*. But by going at such a speed, his chance of hitting the *Pure Oil* would have been much less. That was my attempt to mollify him and give him an answer."

Although he had established a friendship with his fellow pilot Cyrus Epler, Lerro wasn't particularly close to anyone in the association. He and Sophie had few friends in the Tampa Bay area. So he and Hayes—one of his oldest and closest friends—had long, emotional phone conversations several times a week.

Lerro seemed to spend every waking moment replaying the incident in his mind, wondering what he could have done differently. "I said 'You really have to forgive yourself for this thing,'" Hayes recalled. "I said that over and over. But he always felt guilty, guilty, guilty.

"I used to say to him, 'John, it's a damn good thing you didn't become a general in the army. Because if you did, you'd have to send people to their deaths all the time. And you would make blunders that would cause even more deaths, and you'd have to learn from those blunders and go on to become a better general.' But that was very little solace to him, that kind of talk."

Yerrid had a Herculean task before him. He said Lerro always seemed to be playing devil's advocate: "He'd say 'You see, I could have done this, or that! If I'd have done that, then maybe it would've made a difference. . . . ' So I had ten, twenty mini-trials before I ever got to

Fig. 20. Flanked by his lawyer, C. Steven Yerrid, and paralegal Suzanne Reed, John Lerro listens as his fate is decided during the DPR license hearings in October 1980. Courtesy of *Tampa Bay Times.*

the real trial, and I had the harshest juror in the world: I had John. Who wanted to convict more than anybody I knew."

The upcoming "trial" was an administrative procedure in which Lerro would either be branded incompetent and stripped of his license permanently, or walk away fully exonerated. It would almost certainly be the final struggle.

Privately, Hayes was worried.

"The Coast Guard, which is the principal investigating body, is often very reasonable in these situations," he said. "But in my opinion, the pilot associations are concerned that if they don't clean their own house, the pilot commissions are going to come under greater pressure, greater regulatory control, and that might even lead to more regulation of what they can earn. And with all the public outcry, based on my experience as a collision researcher, the attitude was 'give him a fair trial and hang him.'"

Everyone, Yerrid said, "fully expected us to lose. They thought it would be fun to watch the hanging."

* * *

In the weeks following the *Summit Venture* incident, Atkins, the former Gulf Oil captain, started to question whether he'd made the right career choice.

Atkins had moved his wife and two young children to Florida from Massachusetts, the only home the kids had ever known, in hopes that working as a pilot would allow him to spend more time with them, to take on more of the responsibilities of parenting. He was feeling guilty about being away at sea for those four- and six-month stretches. And the marriage was already in trouble; it had taken him a long time to convince Janis Atkins to give things a fresh start in the South.

After the accident, Atkins recalled, "My day to day activities are pilot a ship, go to the law office, try to get some sleep, pilot a ship, go to the law office . . . this is my day. And I'm in a totally new venue that I don't know anything about. It became a little bit overwhelming."

Because the media's focus was on Lerro, Atkins and his family were able to live their lives in relative anonymity. No one really knew what he looked like. Over time, however, piloting Tampa Bay became stressful ("I felt my heart in my throat more times than I ever thought I would, getting around the docks"), and the possibility that he might have another accident down the line weighed heavily on him.

"The other thing that was always ringing in my mind through the months that I stayed down there was a comment from one of the Coast Guard guys. He said 'You just started, and you've got a tough road. Three strikes and you're out down here, and you already got one major one against you.'"

Janis announced her intention to return to Marblehead so that their daughter could start kindergarten in the fall. Bruce was already feeling he didn't belong. The escape hatch was starting to look pretty good.

"I came out of the shipping industry, where you've got thirty guys on a ship and you're all on the same team," he said. "You may be crew members, you may be officers, but if some kind of shit hits the fan

Fig. 21. Bruce Atkins, photographed in 2009. The accident and its aftermath soured him on piloting, and he left Tampa after less than four months for his native Massachusetts. Photo by the author.

you've got to depend upon each other. You're a team out there. And when you go on vacation you may not even talk to anybody until you get back, but you're a team.

"So I naturally carried that forward into this: OK, here's my new team." But the Tampa Bay Pilots Association, according to Atkins, was not supportive. "They didn't behave at all like a new team. They went away and said 'Good luck.' And they stuck their heads in the sand. And so I'm a brand new guy, the Coast Guard's telling me I got a major strike against me, and I'm out there dealing with lawyers and reporters every day. And my teammates aren't giving me any guidance."

He resigned in July 1980, just two months after the accident, and returned to Massachusetts and Gulf Oil, where he had given up his seniority by moving to Florida. The ink was still wet on his Pilot Association contract. "Sue me," he told them. "This isn't what I signed up for."

"I wish there was a reset," Atkins said. "I wish you could catch your breath. I enjoyed the area, I enjoyed doing it, and I enjoyed a number of the people."

He and his family had been befriended by pilot Gary Maddox and his wife. "Gary and Linda were just tremendous people," Atkins said, "and I feel like I let them down. It didn't work. I wish it had. Because I would have made the right choices and decisions, because I think I would have done it for years and enjoyed every breath of it."

He never saw John Lerro again.

* * *

The hearings began on October 20, Lerro's thirty-eighth birthday. Over four days inside a cramped room at the Hillsborough County Courthouse, administrative judge Chris Bentley listened to more than twenty witnesses, including Lerro himself, as prosecutor Steve Oertel laid out the state's case against its beleaguered deputy pilot.

In his opening statements Oertel told Bentley that he intended to prove Lerro alone was responsible for the collision. "The vessel was not blown into the Skyway Bridge," he said. "It was *steered* into the Skyway Bridge."

Oertel had asked the judge to allow transcripts of the Asian crew members' May testimony to be admitted, but Bentley denied the request, saying that the prosecutor had made insufficient efforts to subpoena the men, who were now back on the other side of the world.

In his opening statements Yerrid accused the state of "name calling" and "character assassination," saying that the suggestion that Lerro's mind may have been elsewhere on May 9—on his pending bank loan, perhaps?—had ruined the pilot's reputation.

Yerrid also criticized DPR representatives for implying, in public, that Lerro had had "troubles" in Panama (he didn't) and that might be suffering from a "mental impairment" (he wasn't). "We've waited till now from May 9 to have our day in court," the young lawyer declared, "and now we're going to have it."

DPR's first expert witness, Ernest Clothier, was a retired harbor pilot, a former president of the American Pilots Association and the International Pilots Association, and a wizened veteran who had testified at dozens of administrative proceedings like this one. Prosecutor Oertel asked if Clothier believed Lerro, from the time he turned *Summit Venture* at Buoy 2A, less than a half mile from the Skyway Bridge at slow ahead, had navigated in a "professional and prudent" manner.

Clothier cleared his throat and sat up straight. "I think, under the circumstances, if he was in a position I think he was, that he had to steer without the aid of radar," he said. "In other words, he was steering courses and distances.

"Personally, I have no problem with Capt. Lerro going through the bridge. I think he should have gone around in a more sharper fashion and brought the ship around to the next course immediately, or as soon as possible. He acted negligently and irresponsibly, and his conduct resulted in this maritime casualty."

Unfortunately for Oertel, that was the prosecutorial high-water mark. On cross examination Yerrid got the prosecution's expert to admit sheepishly that he had never piloted Tampa Bay himself and that he did not believe Lerro was mentally impaired *or* incompetent. "I never said that," explained Clothier, indignantly.

The witness also concurred when Yerrid introduced the legal term "in extremis," as it applied to Lerro's situation on May 9. In the maritime world, "in extremis" described an emergency under which different standards of judgment might, and generally did, apply.

Finally, Yerrid delivered the coup de grace: Didn't you, Captain Clothier, reach your opinions after spending hours in a room filled with nautical charts and diagrams?

Yes, replied the witness.

"That's what's called nautical hindsight, isn't it?" Yerrid asked.

Answered Clothier: "It's also called Monday morning quarterbacking."

From that point until the trial was closed and court adjourned four days later, it was all Yerrid's show. "When Clothier looked at me the way that he did, I knew he was going to tell the truth," Yerrid recalled. "And he did so at his own peril, because he'd been pressured beyond anything. He was supposed to give the party line, and he just couldn't do it. And he was eminently qualified to be a pilot."

In a break with standard legal procedure, prosecution and defense witnesses were called to the stand out of sequence, as they became available to testify.

A leading Florida psychiatrist told the court that he'd spent hours talking with Lerro and found him to be exceptionally well balanced. "It is my opinion that up to the point of this tragedy, John Lerro had functioned well in his capacity as a person, a seaman, and as a harbor pilot," said Dr. Robert M. Wooten, a witness for the defense.

"John Lerro is an analytical sort of individual," Wooten explained. "He is a logical thinker and there's no evidence to assume that he was not in the past. He remains a logical thinker."

A clinical psychologist had also examined Lerro, and his independently reached conclusions were identical.

Yerrid called Gary Maddox and Cyrus Epler, who had replaced Wiltshire as co-managers of the Tampa Bay Pilots Association, along with another pilot, George McDonald. All three veteran bay travelers agreed that Lerro had made the only logical choice, given his poor circumstances that morning.

Even the two pilots who were called by the prosecution—Earl Evans and Jack Schiffmacher, who had both been through the storm that morning—testified that, in their opinions, Lerro had acted impeccably in a tight spot.

"I've thought long and hard on that collision," Schiffmacher told the court, after repeating his story about the radio transmission he *definitely* made prior to leaving the channel. "My professional judgment is that the *Summit Venture* could not have been stopped before hitting the bridge."

Anthony Suarez was an internationally recognized hydrodynamics expert who specialized in maritime catastrophes and the effects of wind and water on vessel movement. As the key expert witness for the prosecution, he had plotted *Summit Venture*'s course and concluded that Lerro had made the 18-degree port turn too late. Therefore, he said, the weather—bad though it may have been—played absolutely no role in the accident. It was human error on the part of the pilot, Suarez said flatly.

But Yerrid knew the renowned professor of hydrodynamics had flawed information. In the elaborate eight-foot-long timeline the prosecution had prepared (and which the defense was allowed to examine before the start of trial), the attorney had noticed that the time of the collision was listed as 7:32 a.m.

The officially recorded Coast Guard time was 7:34, a full two minutes later.

Suarez had prepared his damning testimony based on entirely erroneous numbers; everything he had said on the stand was therefore wrong. And Yerrid called him on it. The professor's last statement before Bentley excused him was: "I presumed I was given accurate information. If the collision time was wrong, I didn't make that mistake, counselor. Someone else did."

The prosecution's final witness gathered up his papers, got out of the chair, and walked straight out the courtroom doors, without saying another word.

The last witness, on October 24, was John Lerro. Just as he had done in front of the Marine Board in May, and at the pilot board's probable cause hearing in June, he recounted his transit on *Summit Venture* from the time he and Atkins boarded at 6:30 in the morning. Lerro was on the stand for three hours.

"He was a smart son of a gun," Yerrid said. "He was not a slow guy. He was very pensive, and very deliberate, and he would be solemn before he would talk a lot of times. And then he would say some pretty amazing things.

"I put a lot of witnesses on the stand, but John was in that top tier of people because he was captivating. Not because of his voice, or his tone, but a combination of all those things. And when the judge was looking at him, I knew we had a shot, because the judge did not believe a lot of their defense."

What it all came down to, Yerrid explained in his closing statements, was not Lerro's competence—or lack of it—nor the speed, size, weight, direction, or mechanical operation of *Summit Venture*. Everything on the ship was functioning just as it should, and the pilot had handled what was handed to him with absolute professionalism.

The wild card, Yerrid said, was the weather. During the trial he had produced two hydrodynamics experts, both of whom had performed the same tests and made the same calculations as Suarez, the prosecution witness. But Hilliard Lubin and Lawrence Ward used the correct time of the collision, 7:34 a.m.

Working backward from there, they both concluded that the wind had shifted dramatically, in a matter of seconds, with an increased speed, just as *Summit Venture* was beginning its turn to port into Cut A channel. The ship had crabbed southward, and neither Lerro nor the crew—in those terrible black seconds—knew anything was amiss. The gyrocompass and the course recorder showed only the forward motion of the vessel.

From his conversations with the other pilots on the bay, Lerro knew before he set out that visibility was spotty, and there were squalls in the area. There had been no severe-weather warning that morning from the National Weather Service. All of which was standard issue for Tampa Bay. Lerro had no indication whatsoever that there was a monster storm with his name on it.

Yerrid's defense was this: the Skyway tragedy was the result of an act of God, a hidden "super cell" inside an otherwise standard-issue squall, impossible to forecast but painfully real all the same. To accomplish this, "I had to first find the storm, then I had to document the storm, then I had to prove the storm was as bad as it was."

Yerrid called local TV meteorologist Bill Kowal to the stand. The attorney had learned that Kowal and his staff had just started test-driving an advanced new radar system called Doppler, which among its other uses could distinguish wind strengths visually and display them accurately, and graphically, through a varied palette of representative colors.

In court Kowal introduced a series of color images from the radarscope that showed the storm front racing eastward from the Gulf of Mexico and into Tampa Bay. At 7:32 a.m., a time-stamp photo clearly showed, the center of the beast was directly over the lower bay. Right on top of the Sunshine Skyway Bridge, and right on top of John Lerro.

Kowal was able to demonstrate that while the wind had indeed shifted direction between 7:00 a.m. and a little before 8:00, in less than an hour, it had completely changed from south by southwest to a blow from west and northwest. "As a professional meteorologist, the strange presentation like this on radar would make me believe that some sort of mesoscale small low-pressure development within the line, or within the area of showers and thunderstorms, was taking place."

Kowal further testified that local weather observers had reported gusts of between 40 and 70 miles per hour at 7:30 on the morning of May 9. One person had part of the roof of his house blown off. Today meteorologists use the word "macroburst" to describe the unexpected, rapid movement of rain-cooled air in myriad downward directions under a squall. "They didn't have enough data at that time to say 'These things happen,'" Yerrid said. "But everybody knew that they happened because they got caught in them."

Yerrid had also called Circuit Court Judge Mark McGarry, a St. Petersburg trial judge who was something of an outdoorsman. McGarry and his wife happened to have been inside their 31-foot trailer at the Fort DeSoto campground, on Mullet Key, on the morning of May 9. Understanding the importance of an eyewitness, the judge had sought Yerrid out and offered to testify about the wind conditions on Mullet Key.

"I can tell you that it was the most intense storm I have ever stood out in," McGarry said on the stand, "and I have lived in Florida for 50 years. I have stood out in hurricanes before, and I have never felt wind any stronger than that particular wind." The judge, who was also an airplane pilot, explained that he was familiar with the peculiar habits of air currents as they struck and found ways around solid objects.

He was absolutely certain, too, that although the ferocious wind had begun from a south-southwesterly direction, it had changed. McGarry testified that his Hobie Cat had a wind direction vane on it—and when he went down to look at the bay ("such a series of whitecaps and wind-blown waves . . . I have never seen anything like it before"), the instrument was indicating a northwesterly direction.

"It was strange that it would come with such intensity and for such a length of time," McGarry said, "that it would be blowing in what I would consider very strong gusts, extremely powerful gusts broken by lulls of calm. Very unusual."

An act of God, Yerrid explained in his closer, "means a force of nature occurs that is unexpected, so intense, and so supervening, as to relieve the human actions taken under the overpowering influence of that force." There could be "no finding of negligence whatsoever if there is a finding that this bridge was brought down by an act of God."

* * *

On December 24, two months to the day after Lerro's riveting turn on the witness stand, Bentley ruled on the matter.

"He called me and told me his decision," said Yerrid. "He said 'Where's Captain Lerro?' I said 'Right now, John's at the mall doing some last-minute Christmas shopping,' as he was always prone to do.

"The judge said, 'Well, I want you to find him, and give him an early Christmas present. I want him to know that he's been exonerated of all charges. I'm dismissing the case.'

"Of course, I got emotionally overwhelmed. I said 'Thank you, and God bless you,' and I hung up. I sat there for a while and I thought *wow, that's a big home run. That's huge*. Then I had to track John down. He goes, '*What?! What?!*' I had to tell him like ten times. He says, 'What do you mean?' and I say, 'We won.' That's all I could say to him. And then we started crying."

In his thirteen-page summation, Bentley picked apart every allegation the state had made against Lerro, including its argument that he had violated one of the key Pilot Rules for Inland Waters:

> Every vessel shall, in a fog, mist, falling snow, or heavy rainstorm, go at a moderate speed, having careful regard to the existing circumstances and conditions.

According to the state, the courts had interpreted "moderate speed" to mean that speed at which a vessel may stop in one-half the distance of the existing visibility.

"Visibility being nonexistent, the only way Lerro could have complied with this rule was to bring the ship to a halt," Bentley wrote. "That option was ruled out as a reasonable option by every expert pilot testifying in this case, including Petitioner's primary expert, Clothier.

"Thus, Lerro's failure to adhere to this rule cannot be deemed negligence. Adherence to it was impossible."

It was total and absolute exoneration. The best Christmas gift John Lerro could have hoped for.

"And for the briefest time," said Yerrid, "I think he totally and fully believed that now life was going to be good for him. That he'd turned everything around."

13

A Pound of Flesh

The state was not going to let Lerro off without a fight. Oertel fired off a twenty-two-page written response in mid-January, listing the reasons he believed Bentley was wrong. "The hearing officer has rejected over 100 years of analysis of maritime law in his decision," the state's hired gun angrily wrote.

Oertel, who had once been a hearing officer himself, also claimed that Bentley had either ignored or excluded evidence pointing to Lerro's dereliction of duty. He argued that Bentley did not use the standards of admiralty law, a separate and distinct body of law used exclusively for maritime matters. The judge, in his ruling, had declared that admiralty law did not apply in a state administrative hearing.

The Board of Pilot Commissions scheduled a vote at their March meeting. Their options were to accept Bentley's ruling, and reinstate Lerro's license, or reject the judge and continue to fight.

On February 9, Captain Edward Grace, the presiding officer at the joint U.S. Coast Guard–NTSB Marine Board of Investigation, released his lengthy report and recommendations to the Coast Guard commandant. Among the board's conclusions:

Lerro turning the ship based on "a single, unconfirmed, nondescriptive sighting of a buoy assumed to be 2A" was "the proximate cause of the casualty."

Another factor was the pilot continuing at half-ahead—"an immoderate speed"—when he could not see 2A after passing Buoy 16.

Had he examined the maneuvering characteristics data when arriving on board, Lerro would have known that the vessel could have been stopped at 2A, and the anchors dropped.

Captain Liu, as master of *Summit Venture*, should have expressed his concern to Lerro when the weather deteriorated, rather than relying solely on the abilities and judgment of the pilot. Under maritime law, the captain is free—and somewhat obligated—at any point in the transit to retake command from the pilot, should he believe the pilot's actions, or inactions, may endanger the vessel.

Grace's recommendations included the installation of a motorist warning system on the Skyway Bridge and the construction of those concrete bumpers, or "fenders," that were conspicuous by their absence, leaving the bridge unprotected and vulnerable to a ship strike.

Although the report was indeed critical of Lerro, it was in essence just the final chapter in a lengthy investigation and no real threat to his livelihood. "To be a pilot, you have to have a Coast Guard license in that area," explained Grace. "Once he gets that license, he no longer has to worry about the Coast Guard taking it away from him because the pilot board, by agreement, will take care of that. They'll take care of the discipline involved. And that creates a problem."

* * *

On March 3 the Board of Pilot Commissioners gathered in a conference room at the Host International Hotel in Tampa to decide, once and for all, what was to be done about Lerro.

After more of the usual verbal jousting between Yerrid and Oertel, the six sitting members of the board, all of whom had read the testimony from the October trial, argued about whether they could consider evidence that they believed was not adequately covered in Bentley's summation.

At 4:00 p.m., however, they voted unanimously to accept Lerro's exoneration. Tampa pilot Jack Schiffmacher, a member of the board at that time, had excused himself from this particular task because had already testified, in the hearing phase, his opinion that Lerro was blameless.

Lerro appeared stunned when the board announced its decision. "Everything's so negative," he said. "And then it's positive."

Afterward, as he walked to his car, he hesitated before responding to reporters' shouted questions. He was asked: Did he plan to return to piloting? "It varies from moment to moment," Lerro answered. "It's the only job I've ever loved." In the hotel parking lot, he laughingly warned the reporters who'd followed him to duck behind a group of poles as he backed his vehicle out.

Clearly frustrated at their inability to discipline one of their own, the members of the board said they felt bound by law to accept Bentley's decision. But they were not happy about it.

"It's a sad day in Mudville," said Tampa lawyer Crosby Few, who described Lerro's decision to "shoot for the hole" beneath the Skyway as "playing Russian roulette with only one empty chamber in the gun."

"Regardless of how we may feel," commented William Jackson, a South Florida pilot and the acting chairman of the pilot board, "there's only one choice."

As he was leaving the proceedings, cruise ship official Peter Whelpton told the *St. Petersburg Times* that to him, "it's much more interesting what's not in the transcript of the testimony. Unfortunately most of the facts weren't brought to light."

Looking back, Yerrid said the pilot board was "the most reluctant group of people I have seen forced to follow the law. It was a reluctance that was palpable. You could touch it. You could feel it. They did not want to accept this guy's exoneration. They wanted to find any reason not to, and they couldn't."

* * *

After twenty ride-alongs with other pilots, to refamiliarize himself with the shipping lanes and the bay currents, Lerro would be allowed to return to the only job he'd ever loved. He was also required to pass a physical examination.

With these tasks successfully completed and his state license restored, Lerro became officially eligible to go back to work on Thursday, April 9. That morning his name was added to the rotation board.

In the early evening he reported to the Shell Oil terminal at the Port of Tampa. His first assignment was to pilot the tug *New Park Sunlight,* pushing a 501-foot Gulf barge called *Bulkfleet Texas* across the bay and underneath the broken Skyway. Normally barges didn't require pilots—but it was the captain's first time on Tampa Bay, so he had requested one.

Local news crews caught up with Lerro just before he boarded. "It wasn't good to be away," he told them vaguely. He had learned over the past eleven months to be wary of reporters, even the ones who had written or broadcast sympathetic things about him. His normally trusting nature had eroded to a raw nub.

"I walk to the beat of a different drummer," Lerro said. "I don't say 'Boy, I'm elated.' I can't explain it. It's nothing you could understand."

Yerrid, who'd switched out his usual snappy suit for a T-shirt and jeans to join his client and friend for the historic occasion, quietly reminded Lerro that he should be concentrating on the job at hand rather than rehashing the past. Together they left the reporters at the Shell dock and strode purposefully up the gangway of *New Park Sunlight.*

The transit was uneventful, and around 9:00 p.m.—two and a half hours after he'd taken the barge out of port—Lerro ordered *New Park Sunlight* disengaged from *Bulkfleet Texas.* They were in the channel off Egmont Key. The barge then continued its journey across the Gulf to Texas under its own power.

The NTSB investigative findings were released to the press the next day. The board concluded in its fifty-four-page report that Lerro had contributed to the casualty by his failure to abandon the transit immediately when his visual and radar navigational references were lost. It also cast blame on the National Weather Service for its failure to issue a severe weather warning for mariners.

And the report echoed the Coast Guard recommendations:

> Contributing to the loss of life and to the extensive damage was the lack of a structural pier protection system which could have absorbed some of the impact force or redirected the vessel. Contributing to the loss of life was the lack of a motorist warning system which could have warned the highway vehicle drivers of the danger ahead.

Because it was purely a fact-finding agency, with no disciplinary powers, the NTSB's report had no more impact on John Lerro than did that of the Coast Guard. It had been almost a year since the Skyway had fallen, and he'd already walked through several levels of hell and was back on the job.

Lives, and the bridge, were being rebuilt.

* * *

"We always thought John was a hypochondriac," Bob Thompson chuckled. "If you met John somewhere, he'd point something out on his hand and say 'What do you think this is?' It became a point of humor for a lot of us: You know, 'What's Lerro got today?'"

In midsummer Lerro began to have difficulty climbing the pilot ladders up the vertical sides of ships. He felt weak and somewhat unsteady on his feet. He staggered at odd times. At first the problem was slight, and he tried to ignore it. He told himself it was the brutal and unforgiving Florida sun, or perhaps some residual pain from the knee injury he'd sustained years before in an automobile accident.

But ascending those 35- to 50-foot lengths of knotted rope and staggered step-boards, straight up a flat steel hull, required strength and finely tuned hand-eye coordination. It was a dangerous part of a dangerous job. If you misjudged it, you could tangle and twist in the ladder and come between the heaving ship and the pilot boat as they moved together through the water. At best you would drop into the

Gulf; other scenarios were a broken leg, or paralysis, or a crushing, painful death. It had happened to pilots he knew.

The problem got worse. In the fall Lerro went to a chiropractor, who ran a pen along the bottom of his left foot. Involuntarily, his toes curled upward. A Tampa neurologist performed the same test and then asked Lerro to follow the pen with his eyes. The neurologist didn't like the way his eyes tracked and sent him to another neurologist for additional tests.

He was back in Hillsborough Circuit Court in October, testifying in the civil trial against Greyhound Bus Lines brought by the families of victims Tawanna McLendon and Mel Russell. It was the first in a series of complex civil trials that would drag on for nearly four years. The families argued that bus driver Michael Curtin—who also died in the accident—was driving the Skyway too fast, "negligently and recklessly," when other vehicles had slowed down or pulled off the road entirely to avoid the wicked weather.

Lerro testified that he did not see the passenger bus fall. Other than that, he said, he really didn't have much to add. Reporters in the room noted his weary, haggard appearance. "The worst part, for John, was that he had to look into the eyes of concerned family members," said Hayes. "They were there monitoring what their lawyers were doing. He could see from the expressions on their faces how troubled they were, understandably, about the loss of their loved ones. They were looking for somebody to blame."

In early December Lerro traveled to Gainesville for neurological testing at Shands Teaching Hospital, at the University of Florida. He was diagnosed with multiple sclerosis, a chronic autoimmune disease that attacks and erodes the central nervous system. Although MS is not always fatal, its progress is inevitable and irreversible. There is no cure.

"He didn't mind the crucifixion before the Florida Pilot Commission, or the trial in which he was exonerated," Hayes reflected. "He didn't mind the NTSB report where two out of three members of the panel condemned him. He had no right to answer them back or anything like that. Those things, he took very well. But the civil lawsuits bore down heavily on him. It was the biggest stress he'd ever experienced in his life. He thought that's what brought on the MS."

Back in Tampa Lerro went to the University of South Florida library and devoured every scrap of published information on multiple sclerosis. "As he explained it to me, you can be a carrier of MS for many, many years and not know that you had the disease," said Hayes. "The disease would not manifest itself. Much the way you can have HIV but not full-blown AIDS. Stress frequently brings out the illness."

Stress.

The Tampa Bay Pilots Association announced that Lerro had requested an open-ended leave of absence. "It's tragic," pilot Jim Gallagher told the *Times*. "If you look on it literally, he's on sick leave. But when you've got a disease such as that, I just don't think we could expect to see him back."

Tapping his union contacts at the Port of New York, Hayes arranged for Lerro to go to sea as a ship's officer. "There was a guy there, Henry, who'd been our janitor and I knew he was a guy we could rely on," Hayes recalled. "He knew how to bring these guys and their gear right down to the ship and get them all set up for the voyage, in terms of their quarters being right. He'd do anything you needed so that you could get a good start and not tax yourself physically.

"I'd had to do this for guys with heart conditions, who were trying to squeeze out another year so they could get their pension. Or guys with bad backs. You had to do a lot of climbing as a ship's officer. But at least they got on the ship with a good back."

Lerro quickly discovered that he no longer had the stamina to stand watch from midnight to 4:00 a.m. and again from noon until 4:00 p.m. after grabbing whatever sleep he could. Or for performing the other duties expected of an officer, or climbing the endless ladders, or negotiating the narrow corridors. He insisted on "sitting" his watch from a chair, which the captain found unacceptable.

"In less than three months, he called me from a telephone booth on a dock and described the situation he was in," Hayes said. "It was just too much for him. He was at the point where he knew he couldn't take it anymore."

14

Do You Know Who I Am?

> There was nothing good in my life for a solid year after the accident. When people die, and you did it, your decision was instrumental, that's pretty rough. . . . I felt unworthy.
>
> **John Lerro**

As the money ran out, the Lerros sold, one by one, the Hillsborough County properties they'd bought as rental investments. They kept the Wayne Road house in Odessa, with the little pond out back for John's rowboat.

Sophie was often away at sea for months at a time, which only widened the chasm between them. The marriage was already unraveling before they'd moved to Florida. Now, with John out of work, profoundly miserable from the rapidly advancing MS, and consumed by guilt and self-recrimination over his role in the Skyway tragedy, there was nothing left. Sophie moved out, taking fifteen-year-old Charles with her.

"People only have so much power and strength and resources to fight the demons," said Yerrid. "For him, that was a daily struggle. You couple that with the MS, and I think you have much more of a needy person than your average Joe. Sophie was a very strong person, but I think she wanted to get on with her life. I don't judge people. I didn't live their life. But for John, it was another tangible demonstration that he wasn't good enough. That he'd failed."

The Lerros' son didn't take it well: "They were kind of estranged before the accident," Charles said. "But I don't think it was the bridge. I think it was the fact that I grew up—they probably didn't want to divorce until I was grown up. I think they stayed together for my sake."

In the summer of 1984 Hayes met Lerro in New York for the twentieth reunion of their Maritime College graduating class. The trip had been Hayes's idea; he thought it might get his old shipmate's mind off his recent troubles. During the commencement speech by Admiral Floyd Miller, the school's president, a light bulb went on over Lerro's head. "Hey John," he whispered to Hayes. "Maybe I can ask the admiral for a job?"

To Lerro's delight Miller was receptive to the idea, and after several months of back-and-forth over the phone, he hired the forty-two-year-old alumnus to teach a course called Nautical Science (basic seamanship), and another called Advanced Ship Handling (docking), for two semesters.

"He was a good professor," Miller recalled. "The kids could ask him questions—'What about this? 'What about that?'—and they would learn a lot from him."

Years later Miller would hire another SUNY alum, Joseph Hazelwood, to serve as watch commander on the school's summer cruise. Hazelwood had been the captain on the *Exxon Valdez* tanker when it struck an Alaskan reef in 1989, resulting in a spill of 11 million gallons of crude oil into environmentally sensitive Prince William Sound. At the time he was one of the most reviled maritime men in the world—just as John Lerro had been. "Having those two guys, who had all kinds of trouble between them, I knew kids would get a lot from them," Miller said.

* * *

In January 1985 Lerro reported to Fort Schuyler for work. He had designed a course syllabus and was ready to enter the world of academia. Mostly, he told his friends, he just wanted to feel he was contributing again.

His body was disintegrating, but his sense of humor was intact. "If not for this job, I don't know what I'd do," he joked. "I can't even

Fig. 22. Spring 1985: Lerro aboard *Empire State VI*, during his brief tenure as an instructor at the SUNY Maritime College. Courtesy of *Tampa Bay Times*.

rob banks because you have to be able to run." Although his MS had advanced to the point where he was dragging his left leg when he walked, Lerro was still too proud to use a cane.

On campus he was assigned a compartment belowdecks on *Empire State VI*—the well-worn successor to the vessel on which he had trained in his college days—along with eight or nine other temporary members of the faculty and staff. He was to live in Cabin 18, barely the size of a small walk-in closet, with a low ceiling and a single porthole through which, if he was at all interested, he could see the massive Throg's Neck Bridge arching over the college marina. Next to an even tinier room marked "Slop Sink Locker," Lerro's accommodation—like everything else belowdecks on *Empire State VI*—stank of diesel fuel, grease, and sweat.

"I think back to my life in Tampa," Lerro told *St. Petersburg Times* writer David Finkel, who spent two days on the ship with him. "It was beautiful. I died and went to heaven. Now I have filth. I touch things, and I'm filthy."

There was no hot water on *Empire State VI*, which made showers uncomfortable and therefore infrequent. Because shaving was a dicey proposition, he had let his beard grow thick and unruly. "I'm living like an animal," he said to Finkel, only half kidding.

In his room Lerro kept a radio tuned to a classical music station. He had a small television set and an old humidifier that he never switched off. On the wall a tacked-up blanket blotted out any sunlight attempting to get in through the porthole.

"Am I pissed off and frustrated?" he asked. "There aren't words in the dictionary."

Lerro's left hand was in a cast. He had punched a wall, he explained, out of sheer frustration. "I'm only 42, but I get so tired," he told another reporter from Florida, in a telephone interview. "I have no drive, no energy. Little noises affect me. I feel like just sitting and looking, which is a physical and mental symptom of this disease."

He met his students at 8:00 a.m. on the top deck, after a quick bowl of bran and wheat germ drowned in tepid water from his bathroom sink. Navigating the narrow, labyrinthine passageways of the vessel and its many nearly vertical stairways was a constant challenge as the multiple sclerosis ate away at his muscles.

To get into his cot at night, Lerro had to duck his head, lean against the nausea-blue wall with his good arm, and let himself fall horizontally. This, he said, was "a miserable existence."

On campus the cold, gusty April wind played havoc with his equilibrium. Like a blind man, he knew every stretch of handrail and banister from the cafeteria to the administration building. "I don't get any balance messages from my feet any more," he told Finkel.

There were snow flurries. And because of his broken left hand, he was unable to button his jacket fully; it flapped cruelly in the biting wind. He dragged his leg as if it were smashed and useless. Students stared. "I feel terrible," Lerro confessed. "I look like a drunk."

Everyone at the school, he knew, was aware of his history. The thick manila folder containing the Skyway media reports was one of the most frequently checked-out files in the campus library. People asked him about it. He never flinched or tried to avoid the subject. His honesty and his humility always won them over.

As Admiral Miller had predicted, he was an excellent teacher. "I failed at being a pilot," Lerro told Finkel, "but I'm teaching these kids good stuff." Five years after the defining moment of his life, however, Lerro still felt a tremendous void. His conversation almost always came back to it.

"I spent thousands of hours thinking about that day," he said. "Thousands of hours. Trying to figure out, why me? Because. Why the poor souls who died? Because. In other words, no answers."

Jeffrey Weiss, class of '78, was attending law school in White Plains and making ends meet by teaching early morning classes at his alma mater. He too slept in one of the ship's miniscule cabins. He and Lerro hit it off immediately. "He was really interesting and funny," Weiss said. "I could tell he had a good heart. And for a short time, he was my best friend. There was no doubt about it."

They would talk late into the night, about classical music. About fine art. About ballet, which Lerro discussed with what Weiss thought was incredible knowledge and passion. "And he loved pretty girls, that's for sure," Weiss said. "He was a mensch. A good guy. He just wanted to laugh, and not have any problems."

One of the first things Lerro did was lay out his backstory for Weiss, who did not pass judgment. "He knew I would understand," Weiss recalled.

Lerro was totally consumed by the Skyway incident, John Hayes said. "When he met people, if he talked to them for any length of time, he would say, 'Do you know who I am? Do you know my background?' And he would feel compelled to tell them he was the guy who hit the Skyway Bridge.

"He thought people who knew him for over five minutes were entitled to have that information shared with them. He got some kind of relief from getting that off his chest. To have an honest relationship with people."

Hayes remained Lerro's main contact with the outside world. They talked on the phone often—Lerro had a habit of calling at odd hours, and Hayes, as his friend and trusted confidant, knew when to listen, when to speak up, and when to shut up.

"He'd keep replaying the incident with me, looking for a clue as to what he could have done," Hayes said. "What he might have done. So he could blame himself."

Although Lerro taught aboard *Empire State VI*, he still had occasion to navigate the main campus. This required the negotiation, each way, of a steep recessed stairwell and the wobbly metal gangway that met the deck of the ship. One afternoon, not long after the *Times* reporter had flown home to file his story, Lerro was making his way up the main stairwell, slowly, dragging his foot. Another teacher whose quarters were also aboard *Empire State VI* came up behind him.

"It's getting worse, you know," the man said.

"Pardon me?" Lerro replied.

"Your multiple sclerosis is getting worse. Pretty soon you won't be able to go up these steps, you won't be able to go up the gangway. And you might as well face reality."

Lerro relayed this incident to Hayes during their next phone conversation. He was stung, he told his old friend. "That was just throwing cold water all over him," Hayes recalled. "And as time went on, he did deteriorate more. And he saw the cold reality."

When the seasons changed, and the Maritime College students started getting excited about the annual summer training cruise, Lerro—tired, weak, and indeed getting worse every day—knew his time aboard *Empire State VI* was at an end. He did not attempt to talk Miller into letting him stay on.

"I've enjoyed teaching these kids," he said. "I love them, and they love me. But I live like a pig in a room the size of six phone booths. That's been my life; that's what I've been coming home to. That depresses the hell out of me. And now, I've got no income again."

He flew back to Florida to consider what he ought to do next. "I need a woman's affection," he complained. "But I have no money, no decent job. I have nothing to offer a woman of consequence, and that's the worst thing. That's all I want.

"The number of coincidences in my life beyond the bridge accident is just incredible. I'm unhappy, I'll tell you that. I feel like my hands are tied. I keep wondering: Is this all there is?"

* * *

Lerro had been approached, through his attorney, about selling his story for a quickie book or a network movie-of-the-week. In fact, the pitches were constant, and had been constant since the days just after the accident—and without fail they were batted away. Lerro had been through enough, had been called every name in the book (and some that weren't), and had lost his family, his job, and his health. The last thing he wanted was to be accused of profiting from the catastrophe.

In 1984 as Lerro was preparing for his teaching stint, he and Yerrid were contacted by Tampa radio talk show host John Eastman, who fancied himself a writer. Eastman and his girlfriend had written a treatment for a screenplay about *Summit Venture* and the Skyway Bridge. It was an action story—with a moral.

"When I read their treatment," recalled Yerrid, "it was as if they were there. Some of the things they had to have imagined, but they imagined correctly. And I thought well, if the story's going to be told—and it should be told—then we ought to have a part in telling it. And that's when John and I got involved. We wanted it to be done really right.

"I had to worry about John and his integrity. That was paramount. I didn't need to make a movie. I just needed to make sure that no movie was made that was going to be bad."

* * *

It was Yerrid's idea to center the piece, which they titled *An Act of God,* around his unlikely victory in Judge Bentley's courtroom. The accident was told in flashbacks, as the young lawyer and his guilt-ridden client prepared for their uphill legal battle.

"This is not a disaster movie dealing with death," co-writer Constance May told a reporter. "It's the story of two men—Lerro and Yerrid. It's the idea of manhood and the rites of passage. The men dis-

Fig. 23. With Yerrid at his side, a nervous Lerro addresses the media at the *An Act of God* press conference in 1986. Seated are director Brian Hutton (*left*) and screenwriter John Eastman. The movie was never made. Courtesy of *Tampa Bay Times.*

cover themselves, discover what justice is and what that word means. It isn't the story of victims. It's justice versus absolution."

Eastman, in the same interview, was even more blunt. "This," he said, "is a conflicting, angry, dramatic story. The story doesn't let Lerro off. And when you get to the end, you know you don't own shit. If you leave not thinking of yourself as John Lerro, then I haven't made a movie."

Lerro himself seemed to have bought into the company line. "This is dealing with an angry man, about how miserable my 42 years on this planet turned," he told a reporter, adding that he blamed himself for the "poor choices and errors" he'd made.

"Sure, it may be a little bit of soul-selling, making money on a mistake. But what about the victims' relatives? They sat through court hearings and took the money. They took it."

The script, in his opinion, was "good, but not great." It still needed work.

"I think the movie may piss people off," Lerro said. "If it does, OK. If they come after me, it won't be the first time, and I'll be ready."

* * *

By May 1986 *An Act of God* had gone through more than a dozen rewrites. Arthur Goodale, the DOT engineer who'd blown the whistle on the bridge's structural instability, was added to the list of writers. Hollywood scribe Mann Rubin was hired to give the story a final polish.

In May 1986 Yerrid called a press conference in a downtown Tampa hotel. He was accompanied by Lerro, Eastman, and Hollywood director Brian Hutton (*Kelly's Heroes, Where Eagles Dare*), who was part of the team for *An Act of God* despite the fact that he hadn't made a film since the 1983 Tom Selleck turkey *High Road to China*.

"It's a disaster motion picture, but in addition it's a courtroom drama, and the intensely personal, private agony of an individual," Yerrid told the media. The film, he explained, had a preliminary budget of $10 million. The start of production was announced for early 1987. With luck on their side, *An Act of God* could be in America's theaters by Christmastime.

"The victims, I assure you this, will not be exploited," he said. "But the fact is, it happened. If we could go back to May 8 and un-make the movie, I'd have at least two votes right here." He pointed to himself and then to Lerro, who sat uneasily next to him, beads of sweat running down his bearded face from the hot television lights.

"When John Eastman came to me," Lerro said, "I sensed that here was a sensitive man who would approach it openly and honestly. I realize that nothing can be changed and that some people will never consider changing their attitudes."

However, Lerro added, "What I hope now is that this story will at least make me a human being. In some people's eyes, I stopped being that six years ago."

Yerrid had been busy behind the scenes, schmoozing with investors from both Florida and California. Bob Graham, who had been the governor of Florida in 1980 and would shortly resign from that office to run successfully for the U.S. Senate, joked about making a cameo appearance as Greyhound driver Michael Curtin.

With luck, the Department of Transportation would allow filming on and around the still-standing 1954 Skyway. Although the cable-stayed replacement bridge was scheduled to open shortly, there wasn't money in the budget to demolish the old model, so it would still be standing there, useless.

In the end nothing ever came of *An Act of God*. According to Yerrid, by the middle of 1987 Hutton was out as director and Academy Award winner John Huston (*Treasure of the Sierra Madre, The African Queen*) was in the bag to replace him. But on August 28 the hard-living Hollywood legend died. "I was with his agent, on my way to his house to get the contract signed," Yerrid said. "In fact, I still have the bottle of wine I was taking him. What could have been! And when he died, so did my $6 million worth of commitments here in Florida. Because he was the name director."

At the same time Yerrid's courtroom career was taking off—he would earn a reputation as one of the most dogged and successful trial lawyers in Florida. *An Act of God* became a casualty, just "one of those things," he said. "It would have been a hell of a film—you talk about a courtroom drama of unbelievable proportions!"

But Lerro didn't give it another thought. He was changing his life. His divorce had been granted in March, and he'd enrolled at the University of South Florida with the intention of becoming a counselor. You didn't need a sturdy pair of legs to advise somebody.

He was dating, he was attending church again, and he was talking regularly with several Catholic priests. He was studying transcendental meditation, and reading up on Buddhism, and eating a macrobiotic diet. The counseling idea had come to him during his teaching stint at the Maritime College.

"I discovered I could relate well to the students," he said. "We could even discuss the accident because they were studying to be professionals. When I told them that one mistake could screw up their careers—they listened."

15

Back to the Sky

In the days following the *Summit Venture* disaster, the most pressing issue facing the government of Florida was economic: how do we get traffic flowing again?

Transportation Secretary William Rose announced two possible solutions. The destroyed southbound Skyway could be rebuilt, just as it was, for $37 million; reconstruction would take somewhere around eighteen to twenty months. Rose also said he was reviewing plans to tear both spans down and build a single four-lane bridge, which would bring the Sunshine Skyway up to Interstate standards. The price tag: $112 million, with a construction period of thirty to thirty-six months.

It was at this moment that former bridge engineer Arthur Goodale began talking to the media about the corners cut during construction of the 1971 Skyway and the subsequent pier cracking. The *Tampa Tribune*, among others, published excerpts from DOT's internal correspondence, revealing the department's long-standing private concerns about the cracks. Even before the accident, the Skyway was in pretty bad shape.

Hence in late May, when Rose announced with great fanfare his decision to go with Plan A—rebuilding the same bridge, albeit with a protective series of concrete fenders around the main piers—the reaction was almost as swift and terrible as the storm that had ambushed John Lerro and *Summit Venture*. The Tampa Bay media corps suspected a conspiracy in Tallahassee.

"Without consulting anyone in Pinellas or Manatee Counties, without consulting anyone in the Legislature, without consulting those responsible for the safe passage of ships and without consulting effectively with Congress, the bureaucrats have decided how they are going to rebuild the Skyway," declared a *Times* editorial published May 23, 1980. "They've chosen the quickest and cheapest alternative. That may be a mistake." While the newspaper applauded DOT's decision to add fenders, it suggested the move was "an admission that failure to protect the spans in the past was a tragic misjudgment."

A few days earlier, Representative Elvin Martinez of Tampa had announced that his investigative committee would not attempt to affix blame for the tragedy. That investigation, concluded the *Times*, was "developing into a whitewash" that would fortuitously leave both DOT and the Department of Professional Regulation in the clear: "The manner in which the bureaucrats decided to replace the span suggests that nothing has changed since 35 people died on May 9."

The Florida House unanimously voted to delay any DOT move for at least six months. Chamber of commerce "task forces" were formed in the three directly affected counties—Pinellas, Manatee, and Hillsborough—to encourage further research into the decision. Hillsborough interests supported the construction of a tunnel underneath the bay, just as they had in the 1940s. A tunnel would mean that shipping need not be hindered by wind and weather nor by the dangers posed by a fixed object blocking the main shipping channel. No matter how big the new ships got—and they were getting longer, wider and heavier every year—they would never have to "shoot for the hole."

The tunnel's projected cost, however, was $500 million, which would mean motorists would be charged about $15 per trip in order to pay off the project's gargantuan bill. And construction would take an estimated ten years. And the questions remained about the suitability of the Florida substrate for safely sinking a tunnel. The tunnel idea disappeared.

* * *

The governor never seriously considered the DOT plan to "rebuild" the bridge. "I became convinced, first, that the bridge was not in great shape before it was hit by the ship," Graham said. "There were a lot of structural problems, and the engineers I talked to indicated that they thought there would be substantial rehab of the bridge required within ten years, under any circumstances. So it was a case of putting good money after bad.

"Second, at that point we were still under the umbrella of the Interstate system. By a specific date in the 1990s, that bridge would have come out from eligibility for Interstate funding. So if we were going to do any major work, this was the time to do it—where we would have a substantial amount of the cost paid by the Federal Highway Fund."

Haydon Burns's dream of making the Sunshine Skyway part of Interstate 75—indeed, the reason he'd fast-tracked the additional span back in the 1960s—had not yet become a reality when *Summit Venture* collided with the bridge. Work had begun on bringing the approach bridges and causeways up to Interstate standards, but the events of May 9, 1980, left the project in limbo.

"As I understood it," Graham said, "when a new facility was built by the Interstate, there was a certain number of years, almost analogous to a warranty on a car, that you could still draw down federal Interstate funds for its repair or enhancement. And we were within that window. But we didn't have much time left in that window."

Replacing the broken Skyway with a new bridge, of modern design and at contemporary Interstate and safety standards, meant Florida would receive a new window on that federal improvement money.

Graham implicitly understood, too, that a rebuilt, '50s-style Skyway would always be a reminder of the tragic events of May 9. Along with business and political interests, he had his constituents to consider. They still had to drive over the thing every day.

"And while there was a lot of nostalgia associated with the old bridge," Graham said, "it was an *old bridge*. It was, at that point, thirty-plus years old—plus it was not the most elegant design. And it was a steel bridge over salt water, which is a recipe for a lot of corrosion." The governor was all too aware of the shellacking DOT had taken. "All of that argued not to patch the old bridge, but to build a new one," he said. "So I began talking to people about a different type of bridge, one that would stand up better."

He was most impressed with the ideas presented by Figg and Muller, a Tallahassee-based design firm. Eugene Figg had started his career with Florida's DOT in 1958 and was mentored by chief bridge engineer W. E. "Bill" Dean, the man for whom the Skyway superstructure was named. Figg and Muller's Florida projects included both the Long Key and Seven-Mile bridges.

Jean Muller had more than thirty-five years' experience in the design of prestressed concrete bridges and was responsible for eighteen of the most recent bridges laid out over the Seine River in his native France.

He was a pioneer of the precast concrete segmental bridge—a large-scale jigsaw puzzle in which the bridge is "assembled" onsite—and with a relatively new support design, the cable-stayed bridge. "It had not been used much, if at all, in the United States up till that time," recalled Graham. "But engineers thought it had a lot of promise, particularly since it was basically a concrete bridge in an over-water situation."

In a cable-stayed bridge the cables supporting the segmental bridge deck (elevated roadway) are laced through a "saddle system" in one or two tall towers, called pylons. The towers then bear the load of the deck. In the common traditional cable-suspension bridge, vertical cables are attached to the horizontal main cables to support the roadway load on each side. The main cables are then anchored into the rock at both ends of the structure to absorb the weight.

Muller showed Graham one of his most recent creations, the Brotonne Bridge in Normandy. Constructed in 1977, the cable-stayed structure used a "fan" design—the clusters of high-strength supporting cables, threaded through the tall pylons, splayed out from top to bottom like the sleek cut of a sailboat against the wind. And they ran down the center of the roadway, with two wide traffic lanes on either side. The effect was breezy and open, a far cry from the rigid steel boxiness of the original cantilever Skyway. Graham loved it.

Technically the design was credited to Jacques Combault, who had turned his longtime collaborator Muller's vision into a set of working blueprints. Combault later became France's most honored bridge designer. Not only was the Brotonne structurally sound; it was aesthetically pleasing. And the governor was acutely aware that Tampa Bay's new bridge had to be stunning to look at. It would, after all, become the area's most iconic landmark.

Graham embraced Muller's ideas for the Sunshine Skyway and announced on January 31, 1981, that the new Skyway, a cable-stayed, ultra-sleek design with four Interstate-ready lanes and every safety-related feature conceivable, would take about four years and cost $215 million (the state had recently pocketed a $13 million insurance check for the broken bridge). Both obsolete spans would then be demolished.

The first phase of construction began in 1982.

* * *

In Tampa courtrooms, Skyway-related lawsuits were turning into a cottage industry of their own. At one point in 1982, as the new Skyway was beginning to gestate, forty-three law firms were scrapping over the "old" bridge's carcass.

At a pretrial hearing that April, U.S. District Court judge George Carr ruled that Lerro was negligent for not aborting the transit as soon as the weather began to deteriorate.

But Lerro had already been exonerated, begun his struggle with MS, and quit the profession. Once it was determined that the pilot would not and could not be boiled in oil—he wasn't going to be paying restitution to anybody—the finger pointing began in earnest.

Some blamed Lerro and Graham in equal measure.

Editorialized the *Herald-Tribune* in May: "If, instead of the present Skyway, the proposed new suspension bridge with a wider channel opening had been in existence at the time, a freighter piloted as imprudently as the *Summit Venture* was would still have crashed into it. It blundered that far off course.

"To spend as much as a quarter of a billion dollars for a bigger bridge that Tampa shipping and phosphate industry interests want—but highway-users do not need—is to squander state and federal taxes."

In the first civil case to go to trial, the families of fourteen victims sued Greyhound, claiming Curtin was driving recklessly on May 9 and could have stopped in time and therefore saved the passengers on the bus. The jury didn't agree, and the bus line got off the hook. Greyhound's lawyers then sued the shipowners.

Relatives of the thirty-five persons who lost their lives in the accident filed damage claims, in excess of $200 million, against Hercules Carriers, the Liberian firm listed as the owner of *Summit Venture,* and its parent company, Wah Kwong Shipping and Investments of Hong Kong.

Hercules filed counterclaims against Florida and Greyhound, and brought in the United States as a third-party defendant for allegedly improper weather warnings. The United States counterclaimed against Hercules.

In U.S. District Court, Admiralty Division, Judge Daniel Thomas in 1983 refused to limit Hercules' liability to $14 million, the declared value of *Summit Venture* itself, and found the crew "incompetent." Attorneys for the plaintiffs had flown to Hong Kong and deposed the sailors. This "testimony," Thomas ruled, proved that the crew displayed "blatant negligence" the day of the accident.

Because of Wah Kwong's unwritten policy against interfering with compulsory pilots, ruled the district court, neither Liu nor his chief mate, Chan, understood that they were not relieved of their ultimate responsibility to ensure the safety of the vessel. Instead of merely trusting Lerro, they should have acted when the weather turned ugly.

And, it transpired, two senior members of the crew were working without a current license from either China or Liberia. Hercules, Thomas said in a stinging sixty-five-page decision, must shoulder the blame.

The largest settlement, $1.5 million, went to the family of Tampa food broker Charles Collins. He had been heading to a breakfast meeting in Bradenton in his company car, a light blue 1980 Ford Grenada. About $1.2 million was awarded to Albert "Rusty" Krumm, a twenty-two-year-old New Jersey paraplegic whose mother and full-time caretaker, Delores Eve Smith, died in her silver 1979 Volkswagen Scirocco, along with her husband Robert. The damage awards averaged about $300,000 per victim. In his personal injury lawsuit against the shipping company, sole survivor Wesley MacIntire was given $175,000 for his not inconsiderable troubles.

In 1984 the State of Florida filed suit against Hercules for $23.5 million, which included $17.7 million for the destroyed center span of the southbound Skyway, $3.9 million for removing bridge debris from the shipping channel, $336,000 for construction of the detour directing southbound traffic onto the surviving bridge, $758,000 in lost toll revenues, and $784,000 to cover engineering and inspection costs.

Dewey Villareal, the shipowners' attorney, argued that the state should shoulder some of the blame for designing a bridge without protection from collisions and for being lax in its maintenance. The Skyway, Villareal insisted, suffered from "substantial deterioration of the concrete substructure," requiring "major reconstruction," which the Department of Transportation, as everyone knew, had not performed. Therefore, Hercules' liability in the matter should be limited to $8.7 million.

In a twenty-four-page ruling issued on September 18, 1984, Thomas ordered Hercules to pay the State of Florida $19 million. Hercules ultimately lost its appeal.

* * *

Public suspicion about the cable-stayed bridge reached a fever pitch in 1983, as construction progressed and after local TV reporter Bob Hite took it upon himself to scuba dive (illegally) around the foundation of the main pylons looking for "clues" that something was amiss.

Hite discovered a series of hairline cracks, went on the air to cry foul, and forced DOT into an admission that yes, their inspectors had seen the cracks. However, they had determined that these were "heat hydration cracks," a natural and fully expected occurrence in massive concrete pours, which posed no threat to the structural integrity of the bridge. Graham appointed an engineering review panel anyway, and after six independent groups of engineers inspected the piers and came up with the same answer—the cracks weren't dangerous—the issue went away.

But Floridians, understandably, were nervous about the new Skyway and watched its slow construction carefully. At every turn DOT came under a media microscope. Criticisms about engineering and construction practices were leveled on a weekly basis, and the cost of the project—which eventually exceeded $240 million—was the subject of many harsh editorials. In an unprecedented move the department hired an onsite public relations officer to handle all the Skyway media inquiries.

Still, it didn't all go smoothly. In addition to going way over budget, the project took two years longer than original estimates, three workers died in construction-related accidents, and in mid-1984 a 220-ton precast concrete roadway segment fell into the bay when a launching gantry buckled under its weight.

However, said Graham, "by the time the bridge was far enough along so that people could begin to get a sense of what it was going to be, the tide started to turn. And I think today if you did a poll of the users of the bridge, it would be overwhelmingly supported."

Along with the aesthetics, Graham's Skyway was built with a vertical clearance of 194 feet, significantly higher than the original bridge, to accommodate the tallest conceivable vessels, and with the main shipping channel (the "hole" between the two tower spans) widened to 1,200 feet—an expansion of 50 percent. After considerable input from the Tampa Bay pilots, the bridge was built 1,000 feet to the east of the original. This allowed inbound pilots a much greater distance in which to execute the turn into Cut A channel.

An electronic system was installed to warn motorists of any breaks in the bridge, or of pending danger, and closed-circuit TV cameras allowed monitoring of bridge traffic from a dispatch office at the tollbooth complex.

"The Sunshine Skyway accident was really a turning point in the United States," said bridge engineer Michael Knott. "It illustrated the fact that we had a problem and didn't know what to do about it." There were no engineering standards, no risk assessments.

As project manager for the pier protection system, Knott conducted research revealing that twice as many accidents happened with ships hitting the vulnerable approach piers, as *Summit Venture* had done, rather than the tall main piers bordering the channel. "In the old days," he said, "if a bridge were protected at all, it was usually just the main piers in the channel. The situation exists today where we have approach piers that, if they're even slightly touched by a vessel, then they will collapse."

Knott designed a series of fenders—round concrete-and-stone islands that came to be known as dolphins, 60 feet in diameter and rooted 20 feet into the bay substrate—as ship-deflecting bumpers. Each rose 16 feet above the waterline and had 360-degree navigational lights bolted on top. Most important, each of the thirty-two Skyway dolphins could withstand up to 29.6 million pounds of force.

The two all-important main piers were also protected by elliptical rock islands, built around their bases.

The idea was that no vessel—not an outboard, a cabin cruiser, a phosphate bulker, or a container ship 1,200 feet long—would be able to get close enough to the Sunshine Skyway to hurt it.

The price tag for the entire protective system, including the bumpers, motorist warning systems, closed-circuit TV, sensitive guidance and warning systems for pilots, and electronic "collision detectors" on the piers, was $41 million. "We build entire bridges for less money than it took to protect the new Skyway," Knott exclaimed. "There are no cheap solutions. That's the problem."

* * *

In 1986 Graham ran for a seat in the U.S. Senate, which under Florida's "resign to run" law required him to step down from his position as governor. One of the most trusted and well-liked politicians in the South, Graham easily won his seat in the Senate. He had agreed to resign the governor's office on January 3, 1987—the very day new senators were sworn in. Even if he'd lost the election, he would still have had to leave Tallahassee on that day.

Bob Martinez, meanwhile, had been elected Florida's next governor. But there was an overlap: Graham's term did not officially expire until January 6, so Martinez could not take the oath until that

Fig. 24. The $224 million cable-stayed Sunshine Skyway (*foreground*) opened in April 1987, 1,000 feet east of the old spans. After standing idle for four years, both the original bridges were demolished in 1991, and the approaches at both ends were turned into fishing piers. Courtesy of Florida Archives.

day—which meant that Wayne Mixon, Graham's lieutenant governor since 1978, was sworn as the thirty-ninth governor of Florida on a technicality.

Mixon served for the three days—from January 3, when Graham departed, until January 6, when Martinez took over. Had the dedication ceremony taken place during his window, Mixon might well have been the Charley Johns of his era, officiated at a Skyway ceremony, and left future generations to wonder just who the heck he was. Business as usual in Florida politics.

But that isn't what happened. Newly minted Senator Graham was the keynote speaker at the official Sunshine Skyway dedication ceremony on February 7, 1987. The sky was overcast, and the constant threat of rain kept all the speeches short.

Martinez spoke after State Senator Lawton Chiles from Polk County, who would himself become Florida's governor after defeating incumbent Martinez in 1991. The ceremony began with a moment of silence for those who had lost their lives on the bridge on May 9, 1980. Graham may not have understood the irony when he metaphorically referred to the new Skyway, and its pivotal role in linking the northern and southern counties, as "Hands across the Bay."

In a "Skyway Regatta," twenty-six sporty privately owned caravan watercraft raced from Blackthorn Memorial Park, near the rest areas on the north side, to Egmont Key. From 10:00 a.m. to 4:00 p.m. fifteen thousand pedestrians were allowed to walk across the bridge, which was not yet ready for vehicular traffic.

At 7:00 p.m. DOT threw the light switch, and the Sunshine Skyway—Florida's architectural marvel, 1980s-style—lit up the blackness over Tampa Bay like the nightly fireworks at Walt Disney World. Thousands oohed and aahed, and in that instant the original Skyway flickered, faded, and disappeared from memory.

It took an additional two months to finish applying bright yellow paint to the eighty-four tree-trunk-sized cables, and the Skyway wasn't opened to traffic until April 30. Eighteen years later, in 2005, the center span was officially named the Bob Graham Sunshine Skyway Bridge, in honor of the man who had envisioned the project and seen it through. Graham, retired from politics, said he was deeply touched by the gesture.

But in 1987, it was 1954 all over again. "It's going to be a draw, a big draw," proclaimed Jim Breintenfeld, president of the Pinellas Suncoast Chamber of Commerce. "People will travel that route just to see the bridge. I think it's going to create a small industry there." Added designer Eugene Figg: "When people think of Florida in the future, it may well be Disney World, the Cape, and the Skyway."

The Sunshine Skyway Bridge became an official link of Interstate 275 in 1988, at which time the federal government became legally obligated to foot the bill for 90 percent of the structure's maintenance fees.

16

That Stinking Bridge

Sometimes, I don't think I'm so lucky. Sometimes I wish I hadn't survived. People think it's over. It's past.
It isn't.

Wesley MacIntire

Wes MacIntire survived the Sunshine Skyway accident on May 9, 1980, but it killed him all the same.

At first, it appeared that things would probably go back to normal for the MacIntires. As soon as Wes was released from St. Anthony's Hospital, he insisted on paying a visit to *Summit Venture*, which had been delivered from its temporary berth on Egmont to the Tampa shipyards for repairs.

It was just a week after the collision. Captain Liu and his crew were under subpoena to testify during the Marine Board of Investigation hearings—they weren't going anywhere. In a meeting set up by the Coast Guard, the MacIntires presented the crew with a homemade cake with "thank you," in both Chinese and in English, spelled out in white icing.

The MacIntires were given a personal tour of the ship, conducted by Captain Liu. Wes was photographed in the wheelhouse with the captain, both of them smiling broadly like old friends. He had Betty shoot a picture of him sitting on the edge of the cot in Sick Bay where he'd spent two long hours on May 9. He took a snapshot of the lifeboat

Fig. 25. Just days after his release from the hospital, Wes and Betty MacIntire boarded *Summit Venture* bearing a gift: a cake with "thank you" written in icing, in both Chinese and English, May 1980. Courtesy of the MacIntire family.

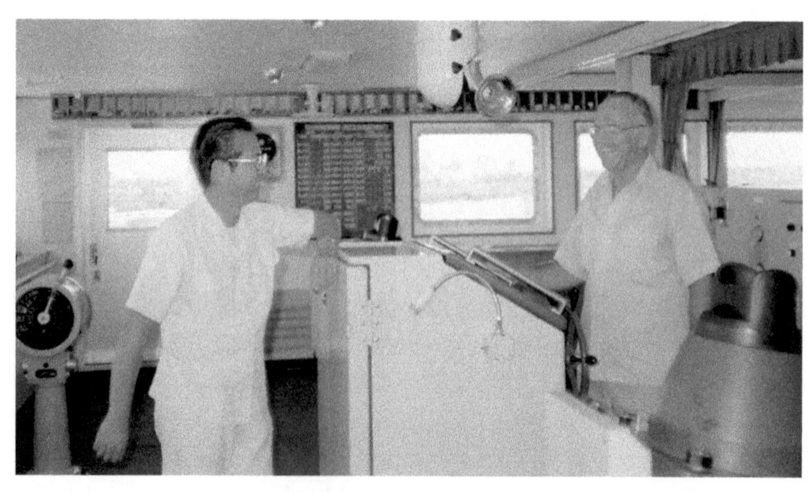

Fig. 26. Wes MacIntire visited Captain Liu Hsiung Chu in the wheelhouse of *Summit Venture*. Wes also had himself photographed sitting on his cot in Sick Bay. Courtesy of the MacIntire family.

Fig. 27. Wes and Betty MacIntire at the Clearwater scrap yard with the remains of the Ford Courier pickup that carried Wes from the top of the Skyway to the bottom of the shipping channel. Courtesy of the MacIntire family.

on which he'd been ferried away from the ship and across the bay to the Fort DeSoto pier. It was back in place on its port side davits and ready to serve once more.

Wes was particularly thrilled to be able to thank Lok Lin Ming, who had heard him screaming from the water on that dreadful morning. The carpenter, who spoke only rudimentary English, grinned shyly as Wes draped an arm around Lok's narrow shoulders for a photograph.

The MacIntires also visited the Clearwater junkyard where the wrecked automobiles, including Wes's beloved Ford Courier pickup, had been dumped after the Sheriff's Department had completed its forensic investigation. Wes and Betty were photographed standing alongside the mangled remains of the little truck that had saved Wes's life.

At the scrapyard owned by Max Zalkin—the man who was selling pieces of the broken Skyway as souvenirs—they picked a small silver ladder out of the wreckage, jammed it into the trunk of Betty's car, and took it home to Gulfport, where it became part of their backyard garden. Over the years, a morning glory vine attached itself and flourished between the rusting rungs.

In July Wes and Betty flew to New York City with Donna and her young son Dean. Wes had been invited to appear on the syndicated revival of the classic TV game show *To Tell the Truth*. The celebrity panel consisted of comedian Nipsey Russell, Canadian actress Tiiu Leek, *WKRP in Cincinnati* star Gordon Jump, and game show perennial Kitty Carlisle. It was their job to hear from three men claiming to be the sole survivor of the Sunshine Skyway Bridge disaster and decide which one was telling the truth.

Three men stood across from them as blow-dried host Robin Ward read Wes's "affidavit":

> I, Wesley MacIntire, am the sole survivor of the Sunshine Skyway Bridge disaster. After being rammed by a freighter, the bridge fell into Florida's Tampa Bay. As I approached it that day in blinding rain and hurricane-force winds, the bridge suddenly swayed. I saw steel beams twist and fall into the water. I couldn't stop. I skidded off the edge and fell 140 feet down into the stormy seas. My falling truck bounced off the ship, which had struck the bridge, and sank. I managed to escape from the cab and was rescued by the crew. Ironically, my life was saved by the very vessel that caused the deaths of 35 other unfortunate people. Signed, Wesley MacIntire.

"Contestant number one" was Wes, wearing a nice new three-piece gray suit and striped tie. He had on a new pair of stylishly tinted eyeglasses. With the heavy TV makeup, his face showed no signed of the injury he'd sustained in May.

The second contestant was another middle-aged man in a sharp suit, about the same age as Wes but with a blonde comb-over instead of a receding hairline. He looked as if he'd never been on the water in his life.

Number three was considerably younger, with unruly black hair, deep hangover eyes and bushy sideburns that gave him a vaguely nautical appearance, like the first mate on a nineteenth-century whaling ship. His collar was undone, his tie loosened.

After a round of vague and clueless questions, gamely answered by Wes and the imposters (who had obviously been given a quick history lesson backstage), Russell and Carlisle correctly identified Wesley MacIntire as Wesley MacIntire, the Skyway guy. They couldn't really explain why he was the one they had believed; in fact, no one seemed to know much of anything about the Skyway tragedy.

The other panelists chose contestant number two, who turned out to be a business consultant. The man with the bushy sideburns was shut out entirely, and everyone laughed when it was revealed that what he did for a living was rent goats to homeowners as organic lawn mowers.

Then came the requisite show business banter. "Wesley," said the unctuous host, "I've heard that this was not your first close call. In fact, you've had several other brushes with death?"

"Yes," Wes replied. His uncomfortable little speech had the stiff sound of repeated rehearsal. "I've had three by ship, and four truck accidents, I fell off a roof once and got hurt, and the big one was off the Sunshine Skyway. Which makes it number nine, and I'm glad I'm not a cat."

Big laughs from the studio audience.

Countered Ward: "Wesley, are you going to be able to make it to your car without assistance?"

"I hope so," Wes said. "I'm going to try very hard."

Cut to commercial.

* * *

Away from the public eye, there was nothing to laugh at. Wes was miserable. It was 1981 before he could bring himself to drive again. Working up the nerve to go over a bridge—any bridge—took even longer.

The nightmares began immediately. The shakes. The sweats.

The oppressive feelings of guilt.

On the first anniversary of the incident, early in the morning, Wes and Betty drove to the Skyway, parked at the Pinellas-side rest area, and walked out to the closed-up western span. Two-lane traffic poked by on the eastern bridge. When the 7:30 Greyhound bus passed, still making its daily southbound route, Wes waved and said a silent

Fig. 28. Every year on the morning of May 9, the anniversary of the accident, Wes MacIntire drove to the Skyway approach. He would wait for the 7:30 bus to Miami—the one he could not save—and wave to it as it went by. Courtesy of the MacIntire family.

prayer. He and Betty dropped thirty-five white carnations over the rail and into the bay, one at a time, one for each victim.

He repeated his ritual visit every year on May 9. It was important to him, he'd say, to pay tribute to those he'd been unable to help back in 1980.

Eventually he felt ready to return to work, driving a truck, but he was so nervous behind the wheel that he quit after a week. A second job, as a crane operator, ended when a cable snapped and mangled two fingers on his left hand.

As the suit against Hercules languished in the court system, Wes and Betty went through their savings—about $10,000—and sold what they could, and borrowed as much as possible.

"I think he just had so much fear, he didn't know where to put it," daughter Donna reflected. "He didn't know what to do. He was scared.

"He'd built a pool in the back of the house. He'd walk around the pool and watch me swim, and I'd say 'Dad, come on in the water.' He'd go 'I don't know if I can.' I had to coax and baby him to get him into the water. That was when it clicked on me how it really affected him.

Until then, I didn't really understand the full impact. It took a lot to pull it out of him. He was scared. He wasn't the same person. Not at all."

Wesley MacIntire suffered from survivor guilt, a form of post-traumatic stress disorder in which the sole survivor of a tragedy feels somehow responsible for not doing enough to save the others. He never got over it.

"I was scared D-Day, but I didn't realize it," he told a reporter in 1985. "I remember great big guys crying. After all I went through in the service, the Skyway seems to be bothering me more than anything. For four years, I felt like I was to blame . . . I'd wake up at night and get mad. I'd swap it. I'd rather go through D-Day again than go over that stinking bridge."

The wrangling attorneys, he said, kept telling him how lucky he was. They called him the Miracle Man. "Sure, I'm bitter," he said. "They said 'You're not hurt.' If I were sitting here with a leg missing, well then you'd say this poor guy got hurt. When you're mentally hurt, people can't see that."

Each summer Wes and Betty spent three months in Brimfield, Massachusetts, staying with Donna and her family. One day Donna came home from work, and Betty said softly, "I want you to go talk to your father."

Donna found him sitting alone in the back yard. "It was the first time I ever saw my father cry," she said. "He goes 'I'm not good any more, I can't do anything. I can't take care of your mother.' He just bawled. "And I said, "Cry, cry, cry. Just get it out.' Because nobody offered physical therapy, or going to see a psychiatrist or anything. And he was a very proud man. He had thought he was going to overcome this, and everything was going to be fine."

In May 1984, four years after he had his standoff with Death on Tampa Bay, Wes accepted a $175,000 settlement from the shipowners. After legal expenses, medical expenses (including the psychiatric counseling he so desperately needed) and his lost wages, he ended up with $75,000.

"What bothered me so much was that John Lerro, within three months, was exonerated," he said. "It takes me . . . years to get a settlement. I had to go to court. And what did I do wrong? I was just driving to work."

Wes never met the man whose fate was inexorably bound with his own. "I sat near Lerro at the hearings, and he never spoke to me," Wes said. "I'd have felt better if he'd talked to me, said he was glad I was alive."

Ironically, he had developed a friendship with Dick Hornbuckle, the man whose Buick sedan had groaned to a stop inches from the edge of the broken bridge. They had started chatting in the courtroom hallways during the interminable hearings and formed an unusual bond—two men who had just barely cheated death. Twice a week Wes and Dick ate lunch together at one St. Petersburg restaurant or another, where they "would talk about everything but the bridge," according to Hornbuckle.

In early 1987, as the finishing touches were being put on the new cable-stayed Sunshine Skyway, the men decided to ask the DOT to let them make the first trip over the new span together in the yellow sedan that Hornbuckle was still driving. "We both lost our money on the toll that day," Wes explained, "so we figured that the state could oblige us this request."

But it didn't happen that way. Notoriously publicity-shy, Hornbuckle opted out, and ultimately Wes accepted the DOT's invitation to close his circle by becoming the last person to drive over the old span before it was officially mothballed. (Neither of the old spans would be dismantled and destroyed until 1991.)

On opening day, April 30, 1987, Wes and Betty arrived in his new blue Ford Courier pickup. It had twin license plates, each with an expression with great significance to Wes—one read "One Day at a Time," the other "Skyway Survivor, 5-9-80."

At 10:00 a.m. the MacIntires led a small procession to the summit of the twenty-three-year-old Skyway, the very span that opera star James Melton had crossed first, back in '54, in his vintage hansom cab—when hopes for Florida and the Tampa Bay area and the sleek, state-of-the-art bridge were as unlimited as the tropical sky.

Wes looked out across the miles of green water toward Tampa and sighed. Then he and Betty took turns letting loose thirty-five long-stemmed white carnations. They watched the flowers flutter and fall 150 feet into the calm water below.

"Thank you God," he said as he climbed behind the wheel of his truck and bowed his head. "I made it."

In the months to come Wes would make numerous trips across the new Skyway and even endorse it for DOT's public relations campaign. "In the beginning, there was a lot of talk about safety of the bridge and questions about the way it was being constructed," he said in the company's propaganda package. "But when I had the opportunity to drive on it, I could feel it was a very safe structure. I have crossed it many times and have no doubt in my mind that it is safe."

But another storm was coming. During the application process for a job driving for UPS, Wes took the mandatory physical exam. The doctors discovered he was in the early stages of prostate cancer. It quickly spread to his bones. Wesley MacIntire died on October 14, 1989, at the age of sixty-five. The family scattered his ashes in Tampa Bay.

Hornbuckle accompanied Betty to the remnants of the old bridge on the tenth anniversary of the accident in 1990. For the last time she followed her husband's ritual, solemnly dropping white carnations into the bay and saying a silent prayer for each of the innocents who died on May 9, 1980.

Her eyes filling with tears behind her dark sunglasses, Betty drew one more flower from the box. She paused for a moment and silently pressed the petals to her lips. Then she let it go—the thirty-sixth flower, for the thirty-sixth victim. Wes was free.

17

The Last Victim

> That's a lot of blood on your hands, even if you didn't do it on purpose. Still, you're on the one who did it. He was, understandably, extremely stressed out by it.
>
> **Bob House**

In the summer of 1988 John Lerro earned a master of arts degree. During one of his infrequent periods of positive thinking, he had enrolled at the University of South Florida for a series of counseling classes through the College of Education.

"I sat back and said, 'Here I am wasting all this human energy getting depressed, thinking that it's a black cloud coming over my head, that it's something I can't change. I realized I can change it, that I'm the guy who's doing it.'"

For two years he traveled nearly every weekday to the sprawling green campus on Fowler Avenue in central Tampa, tenderly extracted himself from his Toyota Camry, and grasping the gold-plated handle of his black walking stick, proceeded slowly and awkwardly to class, his books and papers tucked into the ratty backpack over his shoulder. He might be crippled, but he was determined.

Lerro was not nearly as self-conscious as he had been during his five months at the Maritime College, where the cold and bitter wind played hell with his gait, and where the New York students pretended they weren't staring at him as he stumbled along the footpaths. The

USF campus was flat, like pretty much all the real estate in Florida, and the education building was a straight shot from the parking lot.

His son Charles turned twenty in 1986 and began taking computer classes at the university. Charles, who was also playing guitar, writing songs, and singing in Tampa clubs under the name Chance Lerro, shared a small North Tampa duplex with his dad, near USF. The two frequently had lunch together around the corner at Gus's, a Greek/Italian restaurant that John favored.

John and Sophie's divorce was finalized on March 20, 1987. In the dissolution documents he agreed to return the Haynes flute he'd given her as a gift, and she was ordered to pay her ex-husband $500 per month in alimony. "She had challenged him," Charles recalled. "It was one of those 'So sue me!' things. And he went ahead and did it, and won." The divorce was acrimonious, and this move on Lerro's part ensured that there would never be any sort of truce.

"They weren't friends after that," Charles said. "That's for sure. She really hated him for going after her salary. But she would have hated him anyway."

Before he could work as a counselor, the law required that Lerro serve a local apprenticeship. He volunteered with the Hillsborough County Crisis Center, answering crisis calls two or three nights a week for four hours at a stretch. As a representative of the group he visited area high schools, dispensing advice to kids with home issues. "If somebody comes to me with troubles, it has got to have some relationship to the troubles I've been through," he explained.

He also counseled inmates at a center for convicted criminals on probation. It made him feel like a street kid again—when they cursed, he cursed, like the proverbial sailor. They laughed together and agreed about all the things that sucked in the world. "He felt like a lot of Italians," Lerro's son believed. "They don't feel they can express themselves unless they're being emphatic. So he felt he had a right to be emphatic."

Rather than wallow in depression, Lerro was going to make full use of his unique perspective on failure. "I'd like to be able to explain to the poor bastard who has given up," he said. "I'm hoping someone will come to me with a real big monster and I can make little of it."

Fig. 29. Lerro the counselor. "Everybody screws up," he said. "I know. I invented screwup. When you get caught screwing up, keep your dignity. If you've lost that, you've lost everything." Courtesy of *Tampa Bay Times*.

After the anguish of the early-to-mid 1980s, Lerro's friends and family were glad to see him regain some semblance of a positive outlook. "Everybody screws up," he continued. "I know. I invented screwup. When you get caught screwing up, keep your dignity. If you've lost that, you've lost everything. Self-esteem is the most important thing."

Lerro was prideful, and he hated the thought of being dependent. As his physical condition deteriorated, and his muscles continued their inevitable march to uselessness, he decided he could use his brain to earn a living. "My value now," he said, "is that I can draw upon life's negative experiences and make good of them."

"He always wanted to better himself," reflected Yerrid, to whom Lerro remained close. "He knew he couldn't be a pilot any more. That life was over. So he wanted to do something constructive."

An undated recording exists of one of Lerro's counseling sessions. He is talking with a young woman, probably a high school student. She's chewing gum, and talking a little trash, and the easy flow of their back-and-forth conversation indicates that this was probably not their first session together. Lerro says:

> One of the things I tell suicide people—young kids—when they're mad at their parents and they say "I'm gonna show them. I'll kill myself," I say things like "Why don't you hang around? You're gonna be a bigger pain the ass to your parents if you're hanging around." You're not going to show anybody anything if you commit suicide, 'cause you're gone.

The girl tells Lerro she is going to visit family in New York City. He has a story ready for her:

> I used to go to the Apollo Theatre on 125th Street. That's the heart of Harlem. I'm talking 1962. My roommate and I would be the only white guys there. They had a rock 'n' roll show. And if you'd take a deep breath, you don't have to smoke nothin,' just take a deep breath, you're stoned. That's what that place was like.

After that he's on a roll. He tells her he's been to "every major city in the world," but New York has a unique character all its own. And then he talks for five minutes, uninterrupted, barely taking a breath:

> This is one of my New York sea stories. I was riding down the subway, 8th Avenue and 95th Street, I was going down to 59th and then I was going to walk over to Carnegie Hall. And on the train, right next to me, is standing a girl, college age, maybe 19, 18, 17, pretty little girl. And a guy with her. A very skinny guy. They're carrying books. They're probably freshmen at Columbia University.
>
> In walks a really muscular, weightlifter type. White guy—we're dealing with all whites here—with short hair, a mean-looking guy. And I don't know what happened—in getting into the car the skinny guy might have accidentally touched him, or elbowed him or somethin,' and this big muscle turns to the skinny guy and starts smackin' him! Smacks hurt, you know, a smack in the face.
>
> Of course, the skinny guy, what's he gonna do? This big gorilla weighs about four times what he weighs. He was crying. He was the kind of guy who you know is not going to hit you back.
>
> Well, the girl was crying, saying, "Please leave my boyfriend

alone." This guy was crying. And nobody in the subway car was moving.

I'm standing in the back. And I was in good shape in those days, but I was not a gorilla like this guy was. And I'm thinking to myself, *You can help out if you hit this guy real hard upside the head*—I forget what I was carrying—*but if you hit him too hard, you'll kill him. And if you don't hit him hard enough, he'll kill you.*

At 42nd Street they get off, the big gorilla gets off, and I get off. And I figure well, it's over, we survived that one, everything's OK.

In the middle of every subway station, there's wall-to-wall cops. Well, there's no cops at 42nd Street. The guy starts again! A crowd forms around him as he's smackin' this guy! Everybody's looking, "Where's the cop? Where's the cop?" No cop.

The biggest black man I have ever seen in my life pushes past me. This guy made the gorilla look like nothin'. Me, I'm still trying to decide whether to hit this guy upside the head or not. Because if I hit him, and he don't go down, I'm dead. And I can't run fast enough to get away from this gorilla.

Anyway, this guy—he was an older black guy, with hands about this big—he goes up to the white guy, spins him and says, "Don't you know that hurts?" And the guy looked him up and down—he had to look up—and just walked away. And it ended.

But how about that? That's typical New York. Everybody walked away, and this guy's girlfriend was probably more in love with him than ever. Everything worked out fine.

All kinds of things happen in New York. You get on a subway train, you're liable to hear a guy playing the violin that's just as good as a symphonic violinist.

* * *

"He hated becoming debilitated," said Yerrid. "There were a lot of personal things that it did to him. His ability to be a man. Taking injections and everything to try to stave off the effects. The bottom line was he didn't want to be looked at as something of a cripple, an invalid."

Lerro's son convinced him to terminate the alimony agreement; it was distasteful, he said, that his father continued to demand $500 per month from Sophie. Especially now that Lerro, with his master's degree, would soon start bringing in money as a certified counselor.

And John's father, who had remarried and started a second family in Virginia, had died the year before, leaving a significant amount to John and his older sister Julie. She flew in from Rome for the funeral and was shocked to see her brother's withered limbs. The damage done by the MS, she said, was far worse than anyone could have prepared her for. It was the last time she ever saw him.

Counseling, said Hayes, "was something he could do in an office, sitting down, and as his condition deteriorated he could work out of his home. But by the time he was qualified to work for somebody, he was unable to go to an office, or anything like that. He was not that well established, so he continued doing this counseling for suicide hotlines, crisis hotlines, work for prisoners, out of his home for free. Trying to be socially useful."

But the Hillsborough County Crisis Center let Lerro go, saying his in-your-face, hey-buddy counseling style was just too unorthodox.

"Picture a switchboard with a hundred callers," Yerrid said. "If John was talking to a caller that he believed in, he would let the switchboard go. It didn't matter that they said, 'You can only talk to them for ten minutes.' John would talk to them for hours.

"That's what he told me. He said that he was only governed by his heart in those jobs. He told me this: in trying to save someone else's life, there's no time limits, and no rules can be applied. You do whatever it takes. At whatever cost it requires. That was his philosophy."

Multiple sclerosis was making even the smallest task difficult. "He was getting pretty sick about that time," said Yerrid. "His ambitions never ceased, but it began to take everything for him to get through a day. It drained him. He often couldn't get out of bed."

He began to talk about writing a book, the story of his life. Perhaps, he mused, that would be a way to exorcise the demons that had been perched on his shoulder since that May morning twelve years before. He was beginning to reexplore the Catholic doctrines of his youth, reading books on faith and consulting several priests. "His religion gave him great solace," said Hayes. "And yet he really never let go of the accident."

"I think," Lerro said, "I am going to die not knowing the answers. I look and I say to somebody who knows the answers like my son, who's a Jehovah's Witness, 'You're lucky. You've found an answer that works for you. I don't know any answers."

He knew he was looking at an uphill battle. "There is a reason why I'm here on earth. It wasn't to kill people."

He had documents drawn up creating the John Eugene Lerro Revocable Trust, with himself as trustee. In May 1994 Lerro awarded full ownership of his home, in North Tampa, to the trust. This way, should he become incapacitated, his creditors would not have a legal right to the property. It would go to his heir.

He went out infrequently. It was just too much effort. Still, said Yerrid, "He had a very big awareness of how people perceived him. It took him forever to use a cane. He'd come over to the house for birthday parties and stuff, and he'd sit in a chair. He wouldn't walk around a lot because he didn't want to stumble.

"The bottom line was he didn't want to be looked at as an invalid. Many times we'd be together, and he'd say 'I can walk.' I'd tell him, nobody will know if you lean on me. So he'd put his head on my shoulder and say, 'Just give me a heads up before you stop.' So he'd be putting all his weight on my shoulder as we were walking in somewhere."

Chance Lerro saw his father often in the 1990s, even as the disease broke down his muscles and eroded his mind. Lerro lived off disability checks and the inheritance from his father. "He never lived a lavish lifestyle, ever," Chance said. "He always lived pretty simply. The house that he had moved into was this very small place. He owned that, and he owned his cars, he didn't need a lot of money. When my parents were married, they invested in real estate—rental houses—so there was the money from that."

For the last ten years of his life Lerro was "basically disintegrating," according to his son. "The disease just kept him pretty beaten down. He really hadn't been able to find the kind of activity that was something you could do as a sick person. He was taking classes on writing, and reading a lot of books on that. He was writing down a lot of his old sea stories that used to keep people entertained. They were always the life of the party, him telling those stories."

The stories, ironically, were always from Lerro's days at sea—shipping out as third mate, second mate, chief mate, captain. Crossing the oceans and visiting other continents. He didn't seem to have any good memories of piloting.

On January 10, 1998, Lerro married Roswitha Babette Futrell, a thirty-nine-year-old German woman who had been working as his housekeeper. She was known to her friends as Laila. Several times a week she took her husband to a special swim program. They had handrails installed at the house in Le Clare Shores. Eventually Lerro gave up on the home exercises his doctor prescribed. Just moving his legs caused excruciating pain. Toward the end it took a full-sized electric Hoyer lift—a hydraulic sling used for transporting invalids—to get him out of bed and into his wheelchair.

Chance Lerro was skeptical of his new stepmother's intentions—"I have absolutely no idea if she loved him or not"—but Hayes, who visited often from his home in South Carolina, got along famously with Laila.

As for Yerrid, "I was glad he had someone. It didn't matter who, to me. That was never part of our deal. As long as he was OK, that's all I cared about. The fact that he had a woman—or anyone—that would take the time to care for him. You never know what the motivation was. I didn't go into it because it was unimportant to me. As long as John wasn't alone—because so much of the time, he felt alone, even when there were people in the room."

In November of 2001 Lerro asked Yerrid to take him out on Tampa Bay. "I want to go out there one last time before I die," he said.

And so one sunny afternoon Yerrid helped the grotesquely bent ex-pilot out of his wheelchair and aboard *Justice,* the sixty-five-foot yacht he kept moored at the private dock behind his bayside mansion. They were joined by Chance Lerro and another friend of Yerrid's, former New York Yankee Wade Boggs, who'd recently retired after ending his illustrious career as a member of the Tampa Bay Devil Rays.

As the vessel approached the towering cable-stayed Sunshine Skyway Bridge, it came abeam of 2A, the buoy that still helped direct ship

Fig. 30. John Lerro with Wade Boggs (*left*) and Steve Yerrid, about to make his final voyage on Tampa Bay, November 2001. Courtesy of Yerrid Law Firm.

traffic along the channel and under the bridge. "We were all kind of emotional, talking about life and about how so much had changed," Yerrid recalled. "And all of a sudden we heard this loud blast of a whistle."

Justice was traveling alongside a massive freighter, preparing to make the crucial turn. "The pilot knew who we were. He gave John the salute. I gave John the mic and he said, 'Thanks, Captain Maddox.'

"And the voice came back saying 'No, thank you, Captain Lerro. You made us all better for being here.'

"And John just lost it. I wanted to have a special day for him, but I could not have choreographed that. It was totally out of the blue, and so emotional. John's crying, I'm crying, Wade's crying, my captain's crying."

Near the end of August 2002 John Lerro slipped into a coma. He died on August 31, at the age of fifty-nine.

"He finally quit being haunted by what happened," his widow told the *St. Petersburg Times*. "He finally could stop thinking about it all."

As the end drew near, she added, "I could tell he was looking back on his life. No one deserves to go through what he went through."

Around the world, the obituaries ran with predictable headlines: "John Lerro, Harbor Pilot Haunted by Role in Deadly Bridge Crash," "Pilot in Skyway Disaster Is Dead," "John Lerro, 59, Harbor Pilot in Fla. Crash That Killed 35."

To the *Tampa Tribune*, Laila said: "He spoke to me frequently about it, and he never forgave himself. I hope he found forgiveness for himself in the end. He was an incredibly kind man."

At Lerro's memorial service many of his old friends and shipmates, who hadn't seen one another in ages, reconnected and told story after story about the old days on the high seas. Hayes, House, and Sweeney were there, and Thompson, and many of the Tampa harbor pilots—even those who hadn't thought much of Lerro while he was one of them. Meeting Lerro's son Chance as a man in his mid-thirties, they were struck by his strong resemblance to his dad.

Yerrid gave the eulogy. It was emotional and totally unscripted, Yerrid explained. It came from the heart.

"He knew he was dying. I knew he was dying, and there was nothing I could do. I cared for him greatly. He helped me a lot, as a person, as a human being. He made me understand that life is difficult, and it takes a lot of strength to deal with adversity. And sometimes, despite the best efforts we can put forward, adversity wins."

Postscript

MV *Summit Venture* continued to sail under that name until 1993. Ten years later, after having been bought, sold, refitted and renamed several times, the aging bulker became *Jinmao 9*, added to the considerable fleet of China's Jinmao International, for intra-Asian shipping.

On November 9, 2010, thirty years and six months to the day since the collision with the Sunshine Skyway Bridge, *Jinmao 9*—now flying a Panamanian flag—was en route from Malaysia to China, carrying a cargo of heavy nickel ore, when she was hit by strong winds and 14-foot waves. The ship sank eighty-nine nautical miles east of Cu Lao Cham Island, in Vietnam's Quang Nam Province.

All twenty-seven members of the crew abandoned ship and were rescued without major incident. No salvage was attempted. Written off as a total loss, *Jinmao 9* was left to rust on the bottom of the sea.

Acknowledgments

The author would like to thank all those who agreed to share their thoughts and memories, especially Bruce Atkins, John Hayes, Steve Yerrid, and Chance Lerro.

So many people along the way pulled strings, made introductions, offered stories, made suggestions, and/or helped things move forward, including Roy Adams, Marshall Akers, Gail Atkins, Jane Bartnett, Chris Bentley, Deborah Blum, Jaymi Butler, Jonathan Butler, Chip Burpee, Kris Carson, Anne Cowne, Chris Desa, Karen DeYoung, Shafeek Fazal, Beth Frady, David Finkel, Connie May Fowler, Lauren Hunsberger, Douglas Jordan, Jane Mandelbaum, Jule McGee, William McKeen, Janice Novakowski, Judy Nunez, Stephen Oertel, Bentley Orrick, Suzanne Palmer, Britt Scott, Mitch Stacy, Lee Ann Stiles, Robert Stiles, Paul Tash, Allen Thompson, Roger Vaughn, and Patty Ware. Eternal gratitude to the good people at the St. Petersburg Public Library and the Hillsborough County John F. Germany Library, and to my editors, Sian Hunter, Michele Fiyak-Burkley, Sally Antrobus, and all who sail the UPF.

And finally, to my wonderful, infinitely patient wife, Amy Kagan, who lived the Skyway story with me, over and over again, until I got it right.

Notes on Sources

This book is a work of creative nonfiction. Everything written is based on interviews, court documents, transcripts, newspaper and magazine articles, video and audio recordings, books, letters, and firsthand reporting. It attempts to portray the truth—based on sometimes contradictory reports, explanations, and opinions—about the Sunshine Skyway Bridge, John Lerro, and the *Summit Venture* collision of May 9, 1980. All interviews cited were conducted by the author.

Preface

The Ark lyrics reprinted by permission of Martha Rafferty and the Rafferty family.

Chapter 1. Ambush

The details of the *Summit Venture*/Sunshine Skyway collision were taken from the following:

Testimony: "United States of America Marine Board of Investigation in the Matter of the MV Summit Venture and the Sunshine Skyway Bridge Collision," 1980, testimony of John Lerro, Bruce Atkins, Liu Hsuing Chu, Chan Chim Yee, Sit Hau Po, Lok Lin Ming.

Findings of Fact: Marine Board of Investigation Report and Recommendations to the Commandant, United States Coast Guard, Feb. 9, 1981.

The Accident: National Transportation Safety Board Marine Accident Report: Ramming of the Sunshine Skyway Bridge by the Liberian Bulk Carrier Summit Venture, Tampa Bay, Florida, May 9, 1980, report dated March 3, 1981.

of abutments . . ." Interview with Bruce Atkins, April 21,

ay . . ." Recording of radio call from *Summit Venture*
Guard, May 9, 1980.

across the Bay

ampa Bay Pilots Association . . . egmontkeyferry.com; tampilots.com; Tampa Bay Estuary Program (thep.org); Raymond Arnault, *St. Petersburg and the Florida Dream 1888–1950* (Gainesville: University Press of Florida, 1996), 33.

In Tampa Bay, as throughout the coastal United States . . . *The History of the Tampa Bay Pilots* and *The Pilot Book,* tampabaypilots.com; interview with Gary Maddox, Feb. 22, 2009; interview with Bruce Atkins, April 21, 2009; interview with Judy Nunez, Oct. 12, 2011.

Governor John W. Martin . . . Interview with Bob Graham, Oct. 21, 2010.

In 1924 a trio of enterprising businessmen . . . "Bee Line Ferry speeds service as trade grows," *St. Petersburg Times,* March 20, 1927; "Hopes and frustrations dot Skyway history," *St. Petersburg Times,* Sept. 6, 1954.

The idea of a bridge over lower Tampa Bay . . . "Hopes and frustrations dot Skyway history," *St. Petersburg Times,* Sept. 6, 1954; "One of World's Most Unusual Bridges Steadily Rising in Lower Tampa Bay," *Tampa Tribune,* Jan. 2, 1953; "Skyway span planning began in 1926," *Sarasota Herald-Tribune,* Sept. 2, 1954.

After the war, when the St. Petersburg . . . "Hopes and frustrations dot Skyway history," *St. Petersburg Times,* Sept. 6, 1954; Jon. L. Wilson, "Shaping the Dream: A Survey of Post World War II St. Petersburg," Master's thesis, University of South Florida, 2009.

Florida engineer Freeman Horton . . . *The Road Not Taken: The History of the Sunshine Skyway Bridge,* by Charlie Hunsicker and Allen Horton, tampabay.wateratlas.usf.edu; "Hopes and frustrations dot Skyway history," *St. Petersburg Times,* Sept. 6, 1954.

"That period of growth in the '50s . . ." Interview with Bob Graham, Oct. 21, 2010.

More than twenty thousand suggestions were received . . . "Beach woman named bridge," *St. Petersburg Times,* Sept. 6, 1954; "Sunshine Skyway named by 'attractive housewife,'" *St. Petersburg Times,* July 31, 2002.

At the September 4 dedication ceremony . . . "Skyway a link to peace, prosperity, Van Vleet tells crowd at dedication," *St. Petersburg Times,* Sept, 5, 1954.

September 5 was a Sunday . . . "Bishop hails Skyway as symbol of faith," *St. Petersburg Times,* Sept. 6, 1954.

Across the bay in Manatee County . . . "Manatee celebration events sa Skyway," *St. Petersburg Times*, Sept. 6, 1954.

"The Sun Coast cities have now joined hands . . ." "The hands of good neigh bors and partners reach across the bay," *St. Petersburg Times*, Sept. 6, 1954.

The initial run of . . . "Times will print more Sunshine Skyway papers," *St. Petersburg Times*, Sept. 7, 1954.

Poynter's special edition . . . "Here is guide to your Sunshine Skyway edition," *St. Petersburg Times*, Sept. 6, 1954; "A day for boats, beauties, barbecue," *St. Petersburg Times*, Sept. 7, 1954.

A letter from President Eisenhower . . . "The president is 'delighted,'" *St. Petersburg Times*, Sept. 6, 1954.

Despite the Weather Bureau's prediction . . . "15,086 cars cross over Skyway," *St. Petersburg Times*, Sept. 7, 1954.

Ten coastal counties were represented . . . "Symbolic gestures mark Skyway ceremonies," *St. Petersburg Times*, Sept. 7, 1954.

Rika Diallina, the current Miss Greece . . . nationmaster.com/encyclopedia/Rika-Diallina.

At the precise moment . . . "15,086 cars cross over Skyway," *St. Petersburg Times*, Sept. 7, 1954.

"I present to you . . ." "Beauties symbolize splice of counties in Skyway rites," *St. Petersburg Times*, Sept. 7, 1954.

Fuller Warren, in his speech . . . "Notables, plain folks gather at Palmetto, eat, praise Skyway," *St. Petersburg Times*, Sept. 7, 1954; "Manatee rolls out welcome mat for neighbors to north," *St. Petersburg Times*, Sept. 7, 1954.

Chapter 3. Warning Signs

On Sept. 6, the very same day . . . "U.N. conference told St. Petersburg nation's leading 'retirement city,'" *St. Petersburg Times*, Sept. 7, 1954.

"The fact that there is no monitoring . . ." "History of a Tragedy," *Evening Independent*, Sept. 25, 1980.

Between 1950 and 1960 . . . city-data.com; census.gov; Jon. L. Wilson, "Shaping the Dream: A Survey of Post World War II St. Petersburg," Master's thesis, University of South Florida, 2009.

Disney had quietly begun . . . "Mystery Industry is Disney," *Orlando Sentinel,* Oct. 24, 1965.

Two months later Florida's governor Haydon Burns . . . "Skyway 4-laning plans revealed," *St. Petersburg Times*, Dec. 16, 1965; "Skyway Bridge 4-laning not Interstate must," *St. Petersburg Times*, Jan. 22, 1966; "A Bad Deal," *St. Petersburg Times*, March 1, 1966; "The Skyway Front," *Evening*

ute

larch 1, 1966; "Funds short, Burns will let Kirk build one Petersburg Times, Nov. 22, 1966.

ing Independent . . . "St. Petersburg and the Bridge,"

lowered . . . "Skyway toll: 50 cents in April," *Evening In-* ... b. 12, 1966.

954 blueprints . . . "Experts ignored on Skyway faults" and History Tragedy," *Evening Independent*, Sept. 25, 1980; "Cost came ahead of our safety," *Evening Independent*, Sept. 27, 1980; George Mair, *Bridge Down* (New York: Stein and Day, 1982), 137.

The Florida Department . . . "History of a Tragedy," *Evening Independent*, Sept. 25, 1980.

"As seen from my viewpoint . . ." "History of a Tragedy," *Evening Independent*, Sept. 25, 1980.

"As requested by you . . ." Mair, *Bridge Down,* 141.

The "twin" Skyway was dedicated . . . "Skyway officially opened," *Evening Independent*, May 19, 1971.

Small green-and-white signs . . . "Span dedicated," *Evening Independent*, Nov. 21, 1970.

The 1974 inspection revealed . . . "Bridge cracks said no cause for alarm," *Evening Independent*, Nov. 27, 1974; "Experts ignored on Skyway faults" and "History of a Tragedy," *Evening Independent*, Sept. 25, 1980.

"It would seem there should be . . ." "Skyway scandals," *Sarasota Herald-Tribune*, Oct. 9, 1974.

Every year small boats . . . NTSB Marine Accident Report; "Why didn't the DOT listen?" *Lakeland Ledger*, May 13, 1980.

Then there were the larger issues . . . "What could make the Skyway Bridge fall down?" *St. Petersburg Times*, June 4, 1978.

Concerns about the safety of the twin bridges . . . "What could make the Skyway Bridge fall down?" *St. Petersburg Times*, June 4, 1978.

The crash walls of the tall piers . . . NTSB Marine Accident Report.

"How far should we go . . ." "What could make the Skyway Bridge fall down?" *St. Petersburg Times*, June 4, 1978.

"What has never been officially conceded . . ." "What could make the Skyway Bridge fall down?" *St. Petersburg Times*, June 4, 1978.

In a 1979 memo . . . Mair, *Bridge Down,* 145.

The whistle blower was . . . "Cost-cutting a factor in Skyway collapse, civil engineer says," *St. Petersburg Times*, May 31, 1980; "History of a Tragedy," *Evening Independent*, Sept. 25, 1980.

Chapter 4. The Combat Zone

All details of the collision between *Blackthorn* and *Capricorn* were taken from the following:

"Marine Casualty Report USCGC Blackthorn, SS Capricorn: Collision in Tampa Bay on 28 January 1980," and Recommendations to the Commandant, United States Coast Guard, Dec. 29, 1980.

Judy Kay Nunez, "Blackthorn and Capricorn: Collision with History in Tampa Bay," Master's thesis, Florida State University, 2003.

According to the Port of Tampa . . . "Is Tampa Bay a safe port?" *Evening Independent*, May 16, 1980.

In 1979 the Tampa Bay Pilots . . . "Shipping resumes in bay," *Daytona Beach News-Journal*/Associated Press, May 21, 1980.

On February 9 DOT received . . . Mair, *Bridge Down*, 146.

Shipping ground to a virtual standstill . . . "Pilot is criticized in Skyway crash," *Evening Independent*, Feb. 18, 1980; "Why not fenders?" *St. Petersburg Times*, Feb. 19, 1980; "Pilot involved in 7 incidents before Skyway crash disaster," *Daytona Beach News-Journal*/Associated Press, May 13, 1980.

Chapter 5. Captain Lerro

"All ships move gracefully, slowly . . ." "12 years after Skyway disaster, pilot emerges from shadows," *Miami Herald*, Dec. 20, 1992.

Although he'd been an active member of the Sea Scouts . . . Interview with Julie Lerro, Sept. 13, 2010.

According to family legend . . . E-mail from Chance Lerro, Oct. 3, 2010.

The Lerros believed in art, music, and literature . . . Interview with Chance Lerro, Dec. 28, 2010; interview with Bob House, Sept. 15, 2010.

"They expected me to go to college . . ." "On May 9, 1980, his life changed forever," *Evening Independent*, March 21, 1981.

In August 1960 . . . Interview with John Hayes, Sept. 11, 2010.

Founded in 1874 . . . sunymaritime.edu/MaritimeMuseum/FortSchuyler.

A life at sea was hard . . . Interview with John Hayes, Sept. 11, 2010; interview with Bob Thompson, June 13, 2010.

On campus, cadets were housed . . . Interview with John Hayes, Sept. 11, 2010; interview with Bob House, Sept. 15, 2010.

"He had a romantic, Italian flair . . ." Interview with Bob House, Sept. 15, 2010.

"Meet the right people . . ." Interview with John Hayes, Sept. 11, 2010.

They leapt out of bed every morning . . . Edward Villella with Larry Kaplan, *Prodigal Son* (New York: Simon and Schuster, 1992), 30.

Although Lerro had enrolled . . . E-mail from Chance Lerro, Oct. 3, 2010.

Back in Queens . . . Interview with Bob House, Sept. 15, 2010.

Fort Schuyler was justifiably proud . . . Interview with Bob House, Sept. 15, 2010.

Lerro, perhaps because of his boyhood Sea Scout training . . . Interview with Gene Sweeney, Dec. 11, 2011.

"He had a romantic notion about the sea . . ." Interview with John Hayes, Sept. 11, 2010.

Then there were the ports . . . Interview with Bob House, Sept. 15, 2010.

Back at school, John had discovered . . . Interview with John Hayes, Sept. 11, 2010; interview with Bob House, Sept. 15, 2010.

Villella had also come from . . . Villella and Kaplan, *Prodigal Son*, 28–33.

Lerro began to read everything . . . Interview with Bob House, Sept. 15, 2010.

"You weren't supposed to leave the college during the week . . ." Interview with Bob House, Sept. 15, 2010.

"I was leaving and doing something . . ." . . . Interview with Bob House, Sept. 15, 2010; e-mail from Chance Lerro, Oct. 3, 2010.

One of Lerro's first dance teachers . . . Interview with Bob House, Sept. 15, 2010.

"A friend of mine . . ." Interview with Bob House, Sept. 15, 2010.

In his 1964 senior photo . . . *Eight Bells,* SUNY Maritime College yearbook, 1964.

"We all turned them down flat . . ." Interview with John Hayes, Sept. 11, 2010.

"a big thing to have in your back pocket. If you did get drafted . . ." Interview with Gene Sweeney, Dec. 11, 2011.

"If you took all the mariners out of Brooklyn . . ." Interview with John Hayes, Sept. 11, 2010.

On graduation day . . . Interview with John Hayes, Sept. 11, 2010.

But business was slow on the high seas . . . Interview with Bob House, Sept. 15, 2010.

For seven months, Hayes served . . . E-mail from John Hayes, May 19, 2010.

For a time, he worked for Sea-Land . . . Interview with John Hayes, Sept. 11, 2010.

He obtained his master's license . . . Testimony of John Eugene Lerro to Marine Board of Investigation, May 16, 1980.

"She is the most sensitive, spiritual . . ." "On May 9, 1980, his life changed forever," *Evening Independent*, March 21, 1981.

It was during the six years the family spent in Miami . . . Interview with Chance Lerro, Dec. 6, 2010 and Dec. 28, 2010.

In January 1976 Lerro was hired . . . Testimony of John Eugene Lerro to Marine Board of Investigation, May 16, 1980.

They lived in Coco Solo . . . Interview with Chance Lerro, Dec. 6, 2010.
Sophie taught ballet at the . . . "Ballet classes to start April 18," *St. Petersburg Times*, April 11, 1977.
Young Charles made so many friends . . . E-mail from Chance Lerro, Oct. 3, 2010.
During his time on the canal . . . Testimony of John Eugene Lerro to Marine Board of Investigation, May 16, 1980.
The Florida Legislature . . . Title XXII, Florida Statute 310.011, 1975.
"My dad would literally leave me on the island . . ." Interview with Judy Nunez, Oct. 12, 2011.
"As a pilot, you know all the other pilots . . ." Interview with Bob Thompson, June 13, 2010.
Much like airline pilots . . . Interview with Judy Nunez, Oct. 12, 2011.
"They walked tall . . ." Interview with Judy Nunez, Oct. 12, 2011.
"My father was against the change . . ." Interview with Judy Nunez, Oct. 12, 2011.
"The board shall be composed . . ." Florida Statute 310.011, 1975.
On October 18, 1976 . . . Testimony of John Eugene Lerro to Marine Board of Investigation, May 16, 1980.
The family bought a small . . . "Their paths crossed in disaster," *St. Petersburg Times*, June 22, 1980.
"My father loved all kinds of music . . ." E-mail from Chance Lerro, Oct. 3, 2010.
In 1978 Sophie became one . . . Interview with Chance Lerro, Dec. 28, 2010.
As was the rule . . . Interview with Gary Maddox, Feb. 22, 2009.
"The truth of the matter is . . ." Interview with Judy Nunez, Oct. 12, 2011.
"He was a nice guy, and he had . . ." Interview with Gary Maddox, Feb. 22, 2009.
"He might have been a fine ballet dancer . . ." Interview with Robert Park, Oct. 23, 2010.

Chapter 6. Shoot for the Hole

"Would the accident have happened if . . ." Interview with Bruce Atkins, April 21, 2009.
In the late afternoon of May 8, 1980, Lerro took . . . Testimony of John Eugene Lerro to Marine Board of Investigation, May 16, 1980.
"Because the ship would be going directly . . ." "Retired steamship agent reveals link to ship in Sunshine Skyway collapse," *St. Petersburg Times*, May 7, 2010.

On the ship, Captain . . . Testimony of Liu Hsuing Chu to Marine Board of Investigation, May 13, 1980.

Lerro was to board *Summit Venture* . . . Testimony of John Eugene Lerro to Marine Board of Investigation, May 16, 1980.

"I would have preferred not to . . ." Interview with Bruce Atkins, April 21, 2009.

Lerro radioed *Summit Venture* . . . Testimony of John Eugene Lerro to Marine Board of Investigation, May 16, 1980.

At 6:20 a.m., in near-total darkness . . . NTSB Marine Accident Report.

When they clambered onto the deck . . . Testimony of John Eugene Lerro to Marine Board of Investigation, May 16, 1980.

The wind was light, out of the southwest . . . NTSB Marine Accident Report.

Using a flashlight . . . Testimony of John Eugene Lerro to Marine Board of Investigation, May 16, 1980.

Chapter 7. Tough Old Bird

Wesley MacIntire had a hard head . . . Interview with Betty MacIntire, Dec. 26, 2009.

"As soon as we got inside the harbor we met a French . . ." Letter from Wesley MacIntire to Betty Broadbent, undated, 1942.

On D-Day, June 6, 1944 . . . "Sole survivor: It's a title Wesley MacIntire has earned twice," *Evening Independent*, June 9, 1984.

Back in Boston . . . Interview with Betty MacIntire, Dec. 26, 2009.

"After driving trucks all week long . . ." Interview with Donna (MacIntire) Yeomans, Feb. 12, 2012.

He took a job driving cargo . . . Interview with Betty MacIntire, Dec. 26, 2009.

On Wednesday night, May 7 . . . Mair, *Bridge Down*, 71–87; "The Skyway victims: Varied memories," *St. Petersburg Times*, May 12, 1980.

"Daddy," he said . . . "The day the Skyway fell," *St. Petersburg Times,* May 7, 2000.

A third sister . . . "Relatives of Skyway victims are puzzled over Lerro ruling," *Sarasota Herald-Tribune*, March 3, 1981.

Wes MacIntire stepped out into . . . Interview with Betty MacIntire, Dec. 26, 2009, and followups.

As MacIntire slowed the truck . . . Testimony of Wesley MacIntire to Marine Board of Investigation, June 4, 1980,

So he continued . . . "Sole survivor: It's a title Wesley MacIntire has earned twice," *Evening Independent*, June 9, 1984; Testimony of Wesley MacIntire to Marine Board of Investigation, June 4, 1980.

The next thing he remembered were the words . . . Testimony of Wesley MacIntire to Marine Board of Investigation, June 4, 1980; interview with Betty MacIntire, Dec. 26, 2009.

Chapter 8. The Abyss

"It's a horrible thing . . ." "The Skyway disaster," *Evening Independent*, May 9, 1980.

At 7:00 a.m. Eddie Bartels . . . Eddie Bartels Witness Statement to Donald Blume of the National Transportation Safety Board, May 12, 1980.

Donald Albritton was crawling along . . . Donald Albritton Witness Statement to Jimmie B. Sutton of the National Transportation Safety Board, May 11, 1980.

Ken James had noticed the bus fly past . . . Ken James Witness Statement to Jimmie B. Sutton of the National Transportation Safety Board, May 11, 1980.

Paul G. Hornbuckle had seen . . . Paul G. "Dick" Hornbuckle Witness Statement to Jimmie B. Sutton of the National Transportation Safety Board, May 11, 1980; Testimony of Paul G. "Dick" Hornbuckle to Marine Board of Investigation, May 22, 1980.

Hornbuckle had been known as "Dick" . . . Interview with Lee Ann Stiles (Hornbuckle's niece), June 12, 2010.

Hornbuckle had brought his pal Tony Gattus . . . Interview with Michael Gattus (Tony's son), Sept. 16, 2010.

"I'm hollering to the others . . ." "Memories of Skyway tragedy live on 20 years later," *Ocala Star-Banner*, May 7, 2000.

"I knew some insurance company . . ." "The day the bridge fell," *Evening Independent*, May 7, 1981.

Richard Baserap, on his way to his job as . . . , Richard Baserap Witness Statement to Thomas G. Calderwood of the National Transportation Safety Board, May 11, 1980; Testimony of Richard Baserap to Marine Board of Investigation, May 22, 1980.

Whether by the catching of the anchor . . . Report and Recommendations to the Commandant, U.S. Coast Guard, Feb. 9, 1981.

"*Mayday, mayday, mayday* . . ." Recording of radio call from *Summit Venture* to U.S. Coast Guard, May 9, 1980.

"That," recalled Bruce Atkins, "was very frustrating . . ." Interview with Bruce Atkins, April 21, 2009.

The tugboat *Dixie Progress* was the first to respond . . . Transcript, Testimony to Marine Board of Investigation.

When the dispatcher called again . . . Recording of radio call from *Summit Venture* to U.S. Coast Guard, May 9, 1980.

Chapter 9. The Only Fool Who Went in the Water

"Nobody was ever reviewed . . ." Interview with Robert Park, Oct. 23, 2010.

Wesley MacIntire woke up suddenly . . . Testimony of Wesley MacIntire to Marine Board of Investigation, June 4, 1980; "Sole survivor: It's a title Wesley MacIntire has earned twice," *Evening Independent*, June 9, 1984; interview with Betty MacIntire, Dec. 26, 2009.

Lok shouted something . . . Testimony of Lok Lin Ming to Marine Board of Investigation, May 16, 1980.

Tampa Bay Pilots Association manager B. F. Wiltshire was . . . Interview with Judy Nunez, Oct. 12, 2011, and e-mail followups.

"We knew what his personality was . . ." Interview with Robert Park, Oct. 23, 2010.

The Coast Guard arrived . . . Testimony of Lt. Roy C. Lewis to Marine Board of Investigation, May 16, 1980.

In due course Dewey Villareal . . . Steven Yerrid, *When Justice Prevails* (New York: Yorkville Press 2003), 24.

"That," Atkins said, "was when you wanted to go out on the wing . . ." Interview with Bruce Atkins, April 21, 2009.

Next up the gangway . . . Yerrid, *When Justice Prevails*, 23.

"An eerie quiet . . ." Yerrid, *When Justice Prevails*, 28.

Florida governor Bob Graham happened . . . Interview with Bob Graham, Oct. 21, 2010; *CBS Evening News*, May 9, 1980.

"At that point, human rescue was the focus . . ." Interview with Bob Graham, Oct. 21, 2010.

"It could have happened to any one of us . . ." Interview with Gary Maddox, March 22, 2009.

Under the banner headline . . . "Skyway span falls," *Evening Independent*, May 9, 1980.

Chapter 10. Red Alpha Zulu

The phone rang in Bill Covert's St. Petersburg . . . Interview with Bill Covert, April 20, 2009, and follow-ups by phone and e-mail.

"You always think you're going out . . ." Interview with Mike Rosselet, March 27, 2011.

"One of the first things I saw was a pile of rivets . . ." "The day the Skyway fell," *St. Petersburg Times*, May 7, 2000.

"By this time, Coast Guard was on the scene . . ." Interview with Bill Covert, April 20, 2009, and follow-ups by phone and e-mail.

"The current was so violent . . ." "Vivid memories: Men recall collapse scene, rescue efforts," *Tampa Tribune*, May 9, 2005.

But he was not about to let tow driver . . . "150 feet of terror," *Tampa Tribune*, May 9, 2000.

By 10:30 the two college boats . . . Interview with Bill Covert, April 24, 2009, and follow-ups by phone and e-mail.

A temporary morgue . . . "Everyone agreed, it was the bridge that killed them," *St. Petersburg Times*, May 10, 1980.

Blood from the broken bodies . . . Interview with Bill Covert, April 24, 2009.

A crowd had gathered . . . "Everyone agreed, it was the bridge that killed them," *St. Petersburg Times*, May 10, 1980.

Once the Eckerd team had unloaded . . . Interview with Bill Covert, April 20, 2009.

From his cot in Sick Bay . . . Interview with Betty MacIntire, Dec. 26, 2009.

When he came aboard . . . Interview with Robert Park, Oct. 12, 2010.

Wes MacIntire had been on board . . . Interview with Betty MacIntire, Dec. 26, 2009

Buren told the media . . . "Death rode in on early morning storm," *St. Petersburg Times*, May 10, 1980.

"I knew God was with me on this one . . ." Interview with Betty MacIntire, Dec. 26, 2009.

"He was awake . . ." Interview with Donna (MacIntire) Yeomans, Feb. 22, 2012.

According to the National Transportation Safety Board . . . NTSB Marine Accident Report.

The Greyhound coach was removed . . . "Vivid memories: Men recall collapse scene, rescue efforts," *Tampa Tribune*, May 9, 2005.

Every ten days . . . Mair, *Bridge Down*, 36.

John and Doris Carlson . . . "The Skyway victims: Varied memories," *St. Petersburg Times*, May 12, 1980.

The others were a light green . . . "Sunshine Skyway Accident Report," District Bridge Inspection, Florida Department of Transportation, June 3, 1980.

Chapter 11. The Court of Public Opinion

"How does it feel to have . . ." Interview with Chance Lerro, Oct. 3, 2010.

C. Steven Yerrid, of course . . . Interview with C. Steven Yerrid, March 2, 2012.

"From Tampa Bay, Florida, the revelation today . . ." *NBC Nightly News,* May 10, 1980.

Under media pressure, the ten-member . . . "Police weigh criminal negligence charges in Skyway-ship tragedy," *St. Petersburg Times*, May 11, 1980; "Pilot involved in 7 incidents before Skyway crash disaster," *Daytona Beach News-Journal*/Associated Press, May 13, 1980.

Most failed, however, to mention . . . "Pilot ruled not negligent in 7 previous mishaps," *St. Petersburg Times*, May 13, 1980.

Even more significant . . . "State board has never disciplined a harbor pilot," *St. Petersburg Times*, May 14, 1980.

In the wake of *Summit Venture* . . . Interview with C. Steven Yerrid, March 2, 2012.

"The St. Petersburg highway department had to collapse . . ." *Saturday Night Live*, May 10, 1980.

"We of the Tampa Bay area have been in . . ." "'Saturday Night' Skyway joke in poor taste," *St. Petersburg Times*, May 16, 1980.

The next morning, the original 1954 span . . . "One span of Skyway may open soon, but officials won't say when," *St. Petersburg Times*, May 11, 1980; "Salvage crew raises bus from bay; undamaged Skyway span reopens," *St. Petersburg Times*, May 12, 1980.

The massive section of latticed steel . . . "Damaged Skyway section blasted free of bridge," *St. Petersburg Times*, May 17, 1980.

On Monday a temporary shipping channel . . . "Skyway death toll hits 34," *St. Petersburg Times*, May 14, 1980.

More than $1 million per day . . . "The Skyway: Two more cars found in bay," *Evening Independent*, May 12, 1980.

Governor Bob Graham declared . . . "Graham declares emergency, asks for federal aid," *St. Petersburg Times*, May 15, 1980.

William Rose . . . "Skyway repair to take 1 ½–3 years, cost $37 million to $112 million," *St. Petersburg Times*, May 13, 1980.

Florida House Speaker Hyatt Brown . . . "Florida House panel to probe Skyway tragedy," *St. Petersburg Times*, May 13, 1980.

Committee member Peter Dunbar . . . "Florida House panel to probe Skyway tragedy," *St. Petersburg Times*, May 13, 1980.

"Being a sailor who has been under . . ." "Restricted Tampa Bay channel traffic could resume Sunday," *Evening Independent*, May 12, 1980.

It fell to Hillsborough state attorney . . . "Criminal charges doubted against freighter's pilot," *Palm Beach Post*/Associated Press, June 15, 1980.

"E.J.'s got a population of people . . ." Interview with C. Steven Yerrid, March 2, 2012.

The Coast Guard and the National Transportation . . . Interview with Edward Grace, Jan. 2, 2012.

Two NTSB lawyers . . . Transcript, Testimony to Marine Board of Investigation, May 13, 1980.

The fifty-two-year-old Taiwanese national . . . Testimony of Liu Hsuing Chu to Marine Board of Investigation, May 13, 1980.

Through an interpreter . . . Testimony of Chan Chim Yee to Marine Board of Investigation, May 14, 1980.

Outside the post office building . . . "Body of another Skyway victim found; toll now 35," *St. Petersburg Times*, May 15, 1980.

The attack on his fellow pilot . . . "Skyway Bridge tragedy has pilots on defensive," *Lakeland Ledger*, May 18, 1980.

Chan finished his testimony . . . Transcript, Testimony to Marine Board of Investigation, May 15, 1980.

Bosun Sit Hau Po . . . Testimony of Sit Hau Po to Marine Board of Investigation, May 15, 1980.

The next morning . . . Testimony of Lok Lin Ming to Marine Board of Investigation, May 16, 1980.

After highly technical testimony . . . Transcript, Testimony to Marine Board of Investigation, May 16, 1980.

"There was a terrific explosion in the ship . . ." Testimony of Earl G. Evans to Marine Board of Investigation, May 16, 1980.

John "Jack" Schiffmacher . . . Testimony of John George Schiffmacher to Marine Board of Investigation, May 16, 1980.

Lerro never heard this second broadcast of intent . . . Testimony of John Eugene Lerro to Marine Board of Investigation, May 17, 1980.

Privately, however, pilot association manager . . . Interview with Judy Nunez, Oct. 12, 2011.

"I will never forget, as long as I live . . ." Interview with Judy Nunez, Oct. 12, 2011.

Although Lerro and Atkins both suspected . . . E-mail from Bruce Atkins, Sept. 6, 2011.

"My father always felt that an injustice was done . . ." Interview with Judy Nunez, Oct. 12, 2011.

While the day's drama was playing out . . . "Damaged Skyway section blasted free of bridge," *St. Petersburg Times*, May 17, 1980.

Junk dealer Max Zalkin bought the steel and concrete . . . "A new collector's item: Pieces of the fallen Skyway span," *St. Petersburg Times*, June 27, 1980.

On Friday, May 16 . . . Transcript, Testimony to Marine Board of Investigation, May 16, 1980.

On Saturday morning . . . Transcript, Testimony to Marine Board of Investigation, May 17, 1980.

"There was a lot of talk beforehand . . ." Interview with Edward Grace, Jan. 2, 2012.

Two days later, on Monday . . . Testimony of Bruce R. Atkins to Marine Board of Investigation, May 19, 1980.

"I have never been engulfed by a storm . . ." Testimony of Bruce R. Atkins to Marine Board of Investigation, May 19, 1980.

On Friday, May 30, Summit Venture . . . Interview with Gary Maddox, Feb. 22, 2009.

Chapter 12. Guilty, Guilty, Guilty

A poll conducted by St. Petersburg-based . . . "Poll asks questions on Skyway, political race and refugees," *St. Petersburg Times*, June 18, 1980.

On June 17, the day before . . . "Lerro may be pilot on ship Wednesday," *Evening Independent*, June 17, 1980; "State suspends license of ship pilot Lerro," *Evening Independent*, June 18, 1980.

John Hayes . . . Interview with John Hayes, Sept. 11, 2010, and followup e-mail.

"John looked to me for answers . . ." Interview with John Hayes, Sept. 11, 2010.

Although he had established a friendship . . . Interview with Nancy (Epler) Calfee, Jan. 18, 2012.

Lerro seemed to spend every waking moment . . . Interview with John Hayes, Sept. 11, 2010; interview with C. Steven Yerrid, March 2, 2012.

"I said 'You really have to forgive yourself . . .'" Interview with John Hayes, Sept. 11, 2010.

"He'd say 'You see, I could have done this or that . . .'" Interview with C. Steven Yerrid, March 2, 2012.

"The Coast Guard, which is . . ." Interview with John Hayes, Sept. 11, 2010.

Everyone, Yerrid said . . . Interview with C. Steven Yerrid, March 2, 2012.

In the weeks following the *Summit Venture* . . . Interview with Bruce Atkins, April 22, 2009.

The hearings began on October 20 . . . "Lerro hearing: Press, state accused of 'name calling,'" *Evening Independent*, Oct. 20, 1980.

Oertel had asked the judge . . . "Lerro to testify today," *Evening Independent*, Oct. 22, 1980.

DPR's first expert witness . . . Yerrid, *When Justice Prevails*, 46.

On cross examination Yerrid got the prosecution's . . . Yerrid, *When Justice Prevails*, 47.

"When Clothier looked at me the way he did . . ." Interview with C. Steven Yerrid, March 2, 2012.

A leading Florida psychiatrist . . . Yerrid, *When Justice Prevails*, 48–50.

Yerrid called Gary Maddox and Cyrus Epler . . . Yerrid, *When Justice Prevails*, 81.

Even the two pilots who were called by the prosecution . . . "Pilot: Lerro should have made sharper turn," *Evening Independent*, Oct. 21, 1980.

Anthony Suarez was an internationally recognized . . . Yerrid, *When Justice Prevails*, 68.

The last witness, on October 24 . . . "Lerro on stand: Pilot says strong wind caused ship's collision," *Evening Independent*, Oct. 24, 1980.

"He was a smart son of a gun . . ." Interview with C. Steven Yerrid, March 2, 2012.

What it all came down to . . . Yerrid, *When Justice Prevails*, 77–90.

But Hilliard Lubin and Lawrence Ward used . . . Yerrid, *When Justice Prevails*, 72 (Lubin) and 84 (Ward).

From his conversations with other pilots . . . Testimony of John Eugene Lerro to Marine Board of Investigation, May 17, 1980.

Yerrid's defense was this . . . Yerrid, *When Justice Prevails*, 80–89.

"I had to first find the storm . . ." Interview with C. Steven Yerrid, March 2, 2012.

Yerrid called local TV meteorologist . . . Yerrid, *When Justice Prevails*, 63–66.

"They didn't have enough data at that time . . ." Interview with C. Steven Yerrid, March 2, 2012.

Yerrid had also called . . . Yerrid, *When Justice Prevails*, 50–55.

An act of God . . . Yerrid, *When Justice Prevails*, 80.

On December 24, two months to the day . . . Interview with C. Steven Yerrid, March 2, 2012.

In his thirteen-page summation . . . *State of Florida Division of Administrative Hearings: Department of Professional Regulation vs. John Lerro, Recommended Order*, Jan. 2, 1981.

"And for the briefest time . . ." Interview with C. Steven Yerrid, March 2, 2012.

Chapter 13. A Pound of Flesh

Oertel fired off a twenty-two-page . . . "Don't exonerate Lerro—state agency," *St. Petersburg Times*, Jan. 24, 1981; "Pilot's exoneration disputed by state," *Palm Beach Post*/Associated Press, Jan. 24, 1981.

The judge, in his ruling . . . *State of Florida Division of Administrative Hearings: Department of Professional Regulation vs. John Lerro, Recommended Order*, Jan. 2, 1981.

The Board of Pilot Commissions scheduled . . . "Don't exonerate Lerro—state agency," *St. Petersburg Times*, Jan. 24, 1981.

On Feb. 9, Captain Edward Grace . . . Captain Edward Grace, Report to U.S. Coast Guard Commandant, United States Coast Guard Marine Board of Investigation, M.S. Summit Venture (Liberian Registry); collision with Sunshine Skyway Bridge, Tampa Bay, FL on May 9, 1980 with multiple loss of life, Feb. 9, 1981.

Grace's recommendations . . . Captain Edward Grace, Report to U.S. Coast Guard Commandant, Feb. 9, 1981.

"To be a pilot, you have to have . . ." Interview with Edward Grace, Jan. 2, 2012.

On March 3 the Board of Pilot Comissioners gathered . . . "Decision expected today in Lerro case," *Evening Independent*, March 3, 1981.

After more of the usual verbal jousting . . . Interview with C. Steven Yerrid, March 2, 2012.

At 4:00 p.m., however . . . "Board clears Lerro of wrongdoing," *St. Petersburg Times*, March 4, 1981.

Tampa pilot Jack Schiffmacher . . . "Decision expected today in Lerro case," *Evening Independent*, March 3, 1981.

"Everything's so negative . . ." "Board clears Lerro of wrongdoing," *St. Petersburg Times*, March 4, 1981.

"It's a sad day in Mudville . . ." "Board clears Lerro of wrongdoing," *St. Petersburg Times*, March 4, 1981.

"the most reluctant group of people . . ." Interview with C. Steven Yerrid, March 2, 2012.

After twenty ride-alongs with other pilots . . . "Pilot reportedly returning to work," *Tampa Tribune*, April 8, 1981.

With these tasks successfully completed . . . "Cleared in Skyway tragedy, Lerro pilots tugboat again," *Palm Beach Post*/Associated Press, April 10, 1981.

In the early evening he reported . . . "11 months after Skyway disaster," *Tampa Tribune*, April 10, 1981; "Harbor pilot Lerro guides barge in bay," *Ocala Star-Banner*/Associated Press, April 10, 1981.

The NTSB investigative findings were released . . . "Lerro partly to blame for crash, national agency rules," *St. Petersburg Times*, April 11, 1981.

And the report echoed . . . NTSB Marine Accident Report.

Because it was purely . . . Interview with C. Steven Yerrid, March 2, 2012.

"We always thought John was . . ." Interview with Bob Thompson, June 13, 2010.

In midsummer Lerro began . . . "For Lerro, Skyway nightmare never ends," *St. Petersburg Times*, May 5, 1985.

He was back in Hillsborough Circuit Court . . . "Ship pilot Lerro testifies in Skyway disaster damage suit," *Sarasota Herald-Tribune*, Oct. 27, 1981.

"The worst part, for John . . ." Interview with John Hayes, Sept. 11, 2010.

In early December Lerro traveled . . . "Pilot in Skyway crash has multiple sclerosis," *St. Petersburg Times*, Dec. 24, 1981.

"He didn't mind the crucifixion . . ." Interview with John Hayes, Sept. 11, 2010.

The Tampa Bay Pilots Association . . . "Pilot in Skyway crash has multiple sclerosis," *St. Petersburg Times*, Dec. 24, 1981.

"There was a guy there, Henry . . ." Interview with John Hayes, Sept. 11, 2010.

Chapter 14. Do You Know Who I Am?

"There was nothing good in my life . . ." "The fall—1 lived, 1 took blame; Survivor bitter and ship's pilot lost career, marriage, health," *Orlando Sentinel*, March 22, 1987.

As the money ran out . . . Interview with Chance Lerro, Dec. 28, 2010.

"People only have so much power . . ." Interview with C. Steven Yerrid, March 2, 2012.

"They were kind of estranged before . . ." Interview with Chance Lerro, Dec. 28, 2010.

In the summer of 1984, Hayes met . . . Interview with John Hayes, Sept. 11, 2010.

"He was a good professor . . ." Interview with Floyd H. Miller, Feb. 20, 2012.

Years later Miller would hire . . . "On the water salt's life took unexpected tack," *Newsday*, June 20, 1993.

"Having those two guys . . ." Interview with Floyd H. Miller, Feb. 20, 2012.

In January 1985 Lerro reported . . . Interview with John Hayes, Sept. 11, 2010.

"If not for this job . . ." "For Lerro, Skyway nightmare never ends," *St. Petersburg Times*, May 5, 1985.

On campus he was assigned a compartment . . . "For Lerro, Skyway nightmare never ends," *St. Petersburg Times*, May 5, 1985.

Although his MS had advanced . . . Interview with John Hayes, Sept. 11, 2010.

He was to live in Cabin 18 . . . "For Lerro, Skyway nightmare never ends," *St. Petersburg Times*, May 5, 1985.

"I'm only 42, but I get so tired . . ." "John Lerro is crippled, bitter man," *Evening Independent*, May 4, 1985.

Everyone at the school . . . "John Lerro is crippled, bitter man," *Evening Independent*, May 4, 1985.

"He was really interesting and funny . . ." Interview with Jeffrey Weiss, May 12, 2009.

"When he met people . . ." Interview with John Hayes, Sept. 11, 2010.

"I've enjoyed teaching these kids . . ." "John Lerro is crippled, bitter man," *Evening Independent*, May 4, 1985.

Lerro had been approached . . . Interview with C. Steven Yerrid, March 2, 2012.

"When I read their treatment . . ." Interview with C. Steven Yerrid, March 2, 2012.

"This is not a disaster movie dealing . . ." "Skyway plot: John Lerro, linked with tragedy, is focal point of movie's outline," *Evening Independent*, Sept. 14, 1984.

By May 1986 *An Act of God* . . . "Lerro is ready to tell his side of Skyway story," *Tampa Tribune*, May 8, 1986.

"It's a disaster motion picture . . ." Video recording of press conference, May 6, 1986.

"When John Eastman came to me . . ." "Lerro is ready to tell his side of Skyway story," *Tampa Tribune*, May 8, 1986.

Yerrid had been busy . . . Interview with C. Steven Yerrid, March 2, 2012.

"I was with his agent . . ." Interview with C. Steven Yerrid, March 2, 2012.

Lerro, for his part . . . "Lerro is ready to tell his side of Skyway story," *Tampa Tribune*, May 8, 1986.

"I discovered I could relate well . . ." "Lerro is ready to tell his side of Skyway story," *Tampa Tribune*, May 8, 1986.

Chapter 15. Back to the Sky

Transportation secretary William Rose announced . . . "Skyway repair to take 1 ½–3 years, cost $37 million to $112 million," *St. Petersburg Times*, May 13, 1980; "New bridge may be able to stand loss of support," *St. Petersburg Times*, May 17, 1980.

It was at this moment . . . "Engineer: Bridge doomed by shortcuts, not ship," *Miami Herald*/Associated Press, May 31, 1980.

Hence in late May, when Rose announced . . . "Skyway to be rebuilt as before; supports to have more protection," *St. Petersburg Times*, May 22, 1980.

"Without consulting anyone in Pinellas or Manatee . . ." "Quick, cheaper fix," *St. Petersburg Times*, May 23, 1980.

The Florida House unanimously . . . "Legislators question plan to rebuild Skyway," *St. Petersburg Times*, May 27, 1980; "House calls for six-month delay in bids to reconstruct bridge," *St. Petersburg Times*, May 29, 1980.

Chamber of commerce "task forces" . . . Holly Waggoner, *Skyway to the Sun* (Sarasota: Omnigraphis Publishers, 1988), 24; "As expected, Graham opts for big, 4-lane Skyway," *Sarasota Herald-Tribune*, Feb. 1, 1981.

"I became convinced . . ." Interview with Bob Graham, Oct. 21, 2010.

He was most impressed with the ideas . . . Waggoner, *Skyway to the Sun,* 24.

Jean Muller had more than . . . "Graceful new bridge a triumph," *Orlando Sentinel*, March 22, 1987; "'Visionary' engineer's legacy spans bay area," *St. Petersburg Times*, March 22, 2002.

"It had not been used much . . ." Interview with Bob Graham, Oct. 21, 2010.

In a cable-stayed bridge . . . "Classification and characteristics of cable-stayed bridges," civilengineeringx.com.

Muller showed Graham one of his . . . Interview with Bob Graham, Oct. 21, 2010.

Technically, the design was credited to . . . "High honors for a true innovator: An interview with Jacques Combault," *Florida Engineering Society Journal*, November 2010.

Not only was the Brotonne . . . Waggoner, *Skyway to the Sun,* 24; "The role of Figg & Muller," *Florida Engineering Society Journal*, March 1987.

Graham embraced Muller's ideas . . . "As expected, Graham opts for big, 4-lane Skyway," *Sarasota Herald-Tribune*, Feb. 1, 1981.

In Tampa courtrooms, Skyway-related lawsuits . . . "Lawyers wrangle over damages in Skyway collapse," *Miami Herald*, July 4, 1982.

At a pretrial hearing that April . . . "'Summit Venture' pilot Lerro negligent, federal judge rules," *Evening Independent*, April 23, 1982.

But Lerro had already been . . . Interview with C. Steven Yerrid, March 2, 2012.

"If, instead of the present Skyway . . ." "Skyway fact and fancy," *Sarasota Herald-Tribune*, May 3, 1982.

In the first civil case to go to trial . . . "Ship pilot Lerro testifies in Skyway disaster damage suit," *Sarasota Herald-Tribune*, Oct. 27, 1981; "Investigator says bus last to fall from Skyway," *Evening Independent*, Oct. 27, 1981; "Tragedy in Tampa Bay," *New London (Connecticut) Day*/Knight-Ridder Newspapers, July 11, 1982.

Hercules filed counterclaims . . . "Lawyers wrangle over damages in Skyway collapse," *Miami Herald*, July 4, 1982.

In U.S. District Court, Admiralty Division . . . In the Matter of the Complaint of Hercules Carriers, Inc., for exoneration from or limitation of liability as owner of the M/V Summit Venture, opening statements, Oct. 12, 1982; "Judge: No limit to ship's liability in Skyway disaster," *St. Petersburg Times*, March 17, 1983.

The largest settlement . . . "Skyway victim's family wins $800,000 in lawsuit," *Sarasota Herald-Tribune*, Feb. 16, 1984.

In his personal injury lawsuit . . . "Span tumble survivor settles," *Daytona Beach News-Journal*/Associated Press, May 2, 1984.

In 1984 the State of Florida filed . . . "State seeks millions in Skyway collapse," *Lakeland Ledger*, May 22, 1984.

In a twenty-four-page ruling . . . "Ship owners told to pay for collision," *Miami Herald*, Sept. 19, 1984.

Hercules ultimately lost its appeal . . . In the Matter of the Complaint of Hercules Carriers, Inc., etc., Plaintiff-Appellant . . . , United States Court of Appeals, Eleventh Circuit, Aug. 26, 1985; "State, others can sue in Skyway tragedy, court rules," *Ocala Star-Banner*/Associated Press, Aug. 27, 1985.

Hite discovered a series of hairline cracks . . . Waggoner, *Skyway to the Sun,* 31; "Sunshine Skyway cracks no threat," *Daytona Beach News-Journal*, Nov. 24, 1983; "3 U.S. Congressmen inspect Sunshine Skyway cracks," *Sarasota Herald-Tribune*, Dec. 12, 1983.

In an unprecedented move . . . "Some Skyway reassurance," *Sarasota Herald-Tribune*, June 10, 1984; Waggoner, *Skyway to the Sun,* introduction.

Three workers died in construction-related accidents . . . Waggoner, *Skyway to the Sun,* 58; "Manatee-Pinellas link has 33-year, often tragic history," *Sarasota Herald-Tribune*, April 30, 1987.

In mid-1984 a 220-ton . . . "Skyway construction accident," *Evening Independent*, Aug. 15, 1984.

"By the time the bridge was far enough . . ." Interview with Bob Graham, Oct. 21, 2010.

Along with the aesthetics . . . "A history of challenges," *Florida Engineering Society Journal*, March 1987.

After considerable input . . . "Computer helps ships navigate Skyway safely," *St. Petersburg Times*, May 24, 1984; interview with Gary Maddox, Feb. 22, 2009; "Pilots: Channel safer but slightest slip could mean trouble," *Orlando Sentinel*, March 22, 1987.

"The Sunshine Skyway accident was really . . ." "Interstate bridge struck by barge collapses, killing 14 people," *Professional Mariner*, August–September 2002.

The price tag for the entire . . . "Span's protective shield called best in the world," *Orlando Sentinel*, March 22, 1987; "Controversial bridge starts new era," *Tampa Tribune*, Feb. 2, 1987.

"We build entire bridges for less . . ." "Span's protective shield called best in the world," *Orlando Sentinel*, March 22, 1987.

In 1986 Graham ran . . . http://en.wikipedia.org/wiki/Wayne_Mixson.

Newly minted Senator Graham . . . "Skyway dedication brings out 15,000," *Lakeland Ledger*, Feb. 8, 1987.

"It's going to be a draw . . ." "Officials touting 'Golden Gate of the South,'" *Tampa Tribune*, Feb. 2, 1987.

"When people think of Florida . . ." "Officials touting 'Golden Gate of the South,'" *Tampa Tribune*, Feb. 2, 1987.

Chapter 16. That Stinking Bridge

"Sometimes I don't think . . ." "Life after the fall a struggle for sole survivor," *Tampa Tribune*, Feb. 3, 1987.

At first, it appeared that things . . . Interview with Donna (MacIntire) Yeomans, Feb. 12, 2012.

As soon as Wes was released . . . Interview with Betty MacIntire, Dec. 26, 2009, and phone followups.

The MacIntires also visited the Clearwater junkyard . . . Interview with Donna (MacIntire) Yeomans, Feb. 12, 2012.

At the scrapyard owned by Max Zalkin . . . "Sole survivor: It's a title Wesley MacIntire has earned twice," *Evening Independent*, June 9, 1984.

In July Wes and Betty . . . Interview with Betty MacIntire, Dec. 26, 2009.

The celebrity panel consisted of . . . Video recording, *To Tell the Truth*, Goodson/Todman and Viacom Enterprises, provided by Marshall Akers of ttt-tontheweb.com.

Away from the public eye . . . Interview with Donna (MacIntire) Yeomans, Feb. 12, 2012.

On the first anniversary of the incident . . . Interview with Betty MacIntire, Dec. 26, 2009.

Eventually he felt ready to return to work . . . Interview with Donna (MacIntire) Yeomans, Feb. 12, 2012.

As the suit against Hercules languished . . . "Suit in bridge fall settlement," *New York Times*, May 6, 1984; "Skyway survivor recalls the terror," *St. Petersburg Times*, Feb. 6, 1987.

"I think he just had so much fear . . ." Interview with Donna (MacIntire) Yeomans, Feb. 12, 2012.

Wesley MacIntire suffered from survivor guilt . . . Interview with Betty MacIntire, Dec. 26, 2009; interview with Donna (MacIntire) Yeomans, Feb. 12, 2012.

"I was scared D-Day . . ." "Sole survivor: It's a title Wesley MacIntire has earned twice," *Evening Independent*, June 9, 1984.

"Sure, I'm bitter . . ." "Life after the fall a struggle for sole survivor," *Tampa Tribune*, Feb. 3, 1987.

Each summer Wes and Betty . . . Interview with Donna (MacIntire) Yeomans, Feb. 12, 2012.

"It was the first time I ever saw my father cry . . ." Interview with Donna (MacIntire) Yeomans, Feb. 12, 2012.

In May 1984, four years after he had his standoff with Death . . . "Sole survivor: It's a title Wesley MacIntire has earned twice," *Evening Independent*, June 9, 1984.

"What bothered me so much . . ." "The fall—1 lived, 1 took blame; Survivor bitter and ship's pilot lost career, marriage, health," *Orlando Sentinel*, March 22, 1987.

"I sat near Lerro at the hearings . . ." "Sole survivor: It's a title Wesley MacIntire has earned twice," *Evening Independent*, June 9, 1984.

Ironically, he'd developed a friendship . . . "A bond unbroken," *St. Petersburg Times*, May 6, 1990.

In early 1987, as the finishing touches . . . "2 survivors wish to lead traffic across new bridge," *Miami Herald*, Feb. 11, 1987.

But it didn't happen that way . . . "A golden way across the bay: Disaster survivor says goodbye to the old Skyway," *St. Petersburg Times*, May 1, 1987.

In the months to come . . . Interview with Donna (MacIntire) Yeomans, Feb. 12, 2012.

"In the beginning, there was a lot of talk . . ." Holly Waggoner, *Skyway to the Sun*, 62.

But another storm was coming . . . Interview with Donna (MacIntire) Yeomans, Feb. 12, 2012.

The family scattered his ashes in Tampa Bay . . . "Widow makes last Skyway visit," *St. Petersburg Times*, May 10, 1990.

Hornbuckle accompanied Betty to the old bridge . . . "Widow makes last Skyway visit," *St. Petersburg Times*, May 10, 1990.

Chapter 17. The Last Victim

"That's a lot of blood on your hands . . ." Interview with Bob House, Sept. 15, 2010.

In the summer of 1988 John Lerro . . . "12 years after Skyway disaster, pilot emerges from shadows," *Miami Herald*, Dec. 20, 1992; University of South Florida enrollment records.

"I sat back and said . . ." "The fall—1 lived, 1 took blame; Survivor bitter and ship's pilot lost career, marriage, health," *Orlando Sentinel*, March 22, 1987.

His son Charles turned 20 in 1986 . . . Interview with Chance Lerro, Dec. 28, 2010 and followup by e-mail.

John and Sophie's divorce . . . Hillsborough County Circuit Court record, March 20, 1987.

And she was ordered to pay her ex-husband . . . Interview with Chance Lerro, Dec. 28, 2010.

"She had challenged him . . ." Interview with Chance Lerro, Dec. 28, 2010.

Before he could work as a counselor . . . Interview with John Hayes, Sept. 11, 2010.

"If somebody comes to me with troubles . . ." "Ex-pilot Lerro sets new course," *Tampa Tribune*, Feb. 3, 1987.
He also counseled inmates . . . Interview with John Hayes, Sept. 11, 2010.
"He felt like a lot of Italians . . ." Interview with Chance Lerro, Dec. 28, 2010.
"I'd like to be able to explain . . ." "Ex-pilot Lerro sets new course," *Tampa Tribune*, Feb. 3, 1987.
"Everybody screws up . . ." "John Lerro, 59: Harbor pilot haunted by role in deadly bridge accident," *Los Angeles Times*, Sept. 9, 2002.
"My value now . . ." "Ex-pilot Lerro sets new course," *Tampa Tribune*, Feb. 3, 1987.
"He always wanted to better himself . . ." Interview with C. Steven Yerrid, March 2, 2012.
"One of the things I tell suicide people . . ." Undated audio recording provided by Chance Lerro.
"He hated becoming debilitated . . ." Interview with C. Steven Yerrid, March 2, 2012.
Lerro's son convinced him . . . Interview with Chance Lerro, Dec. 28, 2010.
And John's father . . . Interview with Julie Lerro, Nov. 20, 2010.
She flew in from Rome . . . Interview with Julie Lerro, Nov. 20, 2010.
Counseling, said Hayes, "was something he could do in an office . . ." Interview with John Hayes, Sept. 11, 2010.
But the Hillsborough County Crisis Center . . . "12 years after Skyway disaster, pilot emerges from shadows," *Miami Herald*, Dec. 20, 1992.
"Picture a switchboard with a hundred callers . . ." Interview with C. Steven Yerrid, March 2, 2012.
He began to talk about writing a book . . . "12 years after Skyway disaster, pilot emerges from shadows," *Miami Herald*, Dec. 20, 1992.
"His religion gave him great solace . . ." Interview with John Hayes, Sept. 11, 2010.
"I think," Lerro said, "I am going to die not knowing . . ." "12 years after Skyway disaster, pilot emerges from shadows," *Miami Herald*, Dec. 20, 1992.
He had documents drawn up . . . Warranty Deed to Trustee Under Unrecorded Trust Agreement, Hillsborough County Circuit Court, May 19, 1994.
He went out infrequently . . . Interview with C. Steven Yerrid, March 2, 2012.
Chance Lerro saw his father often in the 1990s . . . Interview with Chance Lerro, Dec. 6, 2010.
On January 10, 1998, Lerro married . . . Florida Marriage Record, Hillsborough Circuit Court, filed Jan. 13, 1998.
Several times a week . . . Interview with John Hayes, Sept. 11, 2010.

Eventually Lerro gave up on . . . Interview with C. Steven Yerrid, March 2, 2012.

Toward the end it took . . . Interview with Bob Thompson, June 13, 2010.

"I have absolutely no idea if . . ." Interview with Chance Lerro, Oct. 3, 2010.

Hayes, who visited often . . . Interview with John Hayes, Sept. 11, 2010.

"I was glad he had someone . . ." Interview with C. Steven Yerrid, March 2, 2012.

In November of 2001 . . . Interview with C. Steven Yerrid, March 2, 2012.

"And the voice came back saying . . ." Interview with C. Steven Yerrid, March 2, 2012; confirmed by Gary Maddox.

Near the end of August . . . "Pilot in Skyway disaster is dead," *St. Petersburg Times*, Sept. 3, 2002.

"He finally quit being haunted by what happened . . ." "Pilot in Skyway disaster is dead," *St. Petersburg Times*, Sept. 3, 2002.

"He spoke to me frequently . . ." "Summit Venture pilot dies after long illness," *Tampa Tribune*, Sept. 4, 2002.

At Lerro's memorial service . . . Interview with John Hayes, Sept. 11, 2010; interview with Bob House, Sept. 15, 2010; interview with C. Steven Yerrid, March 2, 2012.

Meeting Lerro's son Chance . . . Interview with Gene Sweeney, Dec. 11, 2011; Interview with Bob House, Sept. 15, 2010.

"He knew he was dying . . ." Interview with C. Steven Yerrid, March 2, 2012.

Postscript

MV *Summit Venture* continued to sail . . . "Chinese ship crew rescued in Da Nang waters," en.baomi.com, Nov. 12, 2010; "Jinmao bulker sinks off Vietnam," lloydslist.com, Nov. 12, 2010.

Index

Page numbers in italics refer to illustrations

Bill DeYoung is a native of St. Petersburg, Florida. Nationally recognized for his music journalism, he was a writer and editor at newspapers in Florida and Georgia for three decades.

* * *

The University Press of Florida is the scholarly publishing agency for the State University System of Florida, comprising Florida A&M University, Florida Atlantic University, Florida Gulf Coast University, Florida International University, Florida State University, New College of Florida, University of Central Florida, University of Florida, University of North Florida, University of South Florida, and University of West Florida.

Lightning Source UK Ltd.
Milton Keynes UK
UKHW010152050522
402495UK00001B/45